Nuclear Economy 1

SCIENCES

Energy, Field Directors – Alain Dollet and Pascal Brault

Nuclear Energy, Subject Head – Sylvain David

Nuclear Economy 1

Nuclear Fuel Cycle Economic Analysis

Coordinated by
Jacques Percebois
Nicolas Thiollière

ISTE

WILEY

First published 2023 in Great Britain and the United States by ISTE Ltd and John Wiley & Sons, Inc.

ISTE Ltd
27-37 St George's Road
London SW19 4EU
UK

www.iste.co.uk

John Wiley & Sons, Inc.
111 River Street
Hoboken, NJ 07030
USA

www.wiley.com

Library of Congress Control Number: 2023931851

British Library Cataloguing-in-Publication Data
A CIP record for this book is available from the British Library
ISBN 978-1-78945-094-1

ERC code:
PE2 Fundamental Constituents of Matter
PE2_3 Nuclear physics
PE8 Products and Processes Engineering
PE8_6 Energy processes engineering
SH1 Individuals, Markets and Organisations
SH1_12 Agricultural economics; energy economics; environmental economics

Contents

Preface

Sylvain DAVID[1], Jacques PERCEBOIS[2] and Nicolas THIOLLIÈRE[3]

[1] *IJCLab Orsay, CNRS, IN2P3, France*

[2] *CREDEN, CNRS ART-Dev, Université de Montpellier, France*

[3] *Subatech, IMT Atlantique, Nantes, France*

Nuclear energy initiates passionate debates that are often characterized by radical positions. Nevertheless, this subject benefits from robust core knowledge drawing on many disciplines in the human and social sciences, the technological sciences and even life sciences.

However, the dynamism of this field of research cannot obscure the fact that nuclear energy is a controversial source of energy. Research works having addressed nuclear safety, nuclear waste management and decommissioning issues, as well as those discussing the economic aspects of nuclear energy, have not resulted in a consensual vision, shared by all the actors involved.

The economic aspects of nuclear energy are relatively complex, which is evidenced by the size of this work. The authors wish to explore this issue in an academic and detailed way, describing its complexity while making it accessible.

Future strategies involving the use of nuclear energy are highly heterogeneous around the world. Since the characteristics of the energy and electricity mixes fall within the domain of national sovereignty, each country is likely to develop a specific strategy. Many determinants come into play, historical, socioeconomic or even political.

One of the specificities of nuclear energy is that it engages expenses and revenues over prolonged periods of time: around a decade for the construction of sites, half a century for the operation of a plant and a century or more for waste management. This has led to the development of cost calculation methodologies capable of taking those time frames into account. Even assuming that all costs and benefits are accurately known beforehand, there is no predefined law for comparing costs over such long periods, which is why different methods are available. Due to the uncertainty of the future technical and scientific context, the advancement of knowledge, the social model at work and even geopolitics, it is impossible to predict the effect of a financial flow over 50 years. It is necessary to develop sensitivity analyses capable of inspiring decision-makers and guiding the choices to be made.

At present, the implementation of nuclear energy is mainly destined to the production of electricity. Nuclear power plants, as well as all the other sources of electricity, are connected to an electricity network that must ensure the balance between supply and demand at all times. The increasing popularity of intermittent renewable energies in the context of the energy transition is provoking disturbances in the normal flow of electricity markets, which require a thorough understanding so as to better anticipate energetic needs.

This book addresses all of these questions, with a detailed focus on the specificities and calculation methods associated with activities upstream and downstream of the electronuclear energy cycle. Cost estimation is analyzed, with an emphasis on the management of mining waste. We also study the way in which nuclear energy fits into a larger mix and responds to the rules of the energy market in an ever-evolving socioeconomic context.

For practical reasons, this book is divided into two volumes.

The first volume includes the following four chapters and appendix:

- Chapter 1: The Evolution of Nuclear Energy in the World and in France;

- Chapter 2: Nuclear Costs: Methodological Aspects;

- Chapter 3: The Production Costs of Nuclear Electricity;

- Chapter 4: The Costs of Nuclear Fuel;

- Appendix: Ore Deposits and Mining Projects.

The second volume includes the following five chapters and appendix:

- Chapter 1: Management of Spent Fuels;

– Chapter 2: Nuclear Power in the European Electricity Market;

– Chapter 3: The Industrial Challenges of Nuclear Power;

– Chapter 4: The Economic Costs of a Nuclear Accident;

– Chapter 5: Prospective Scenarios from the Present to 2050;

– Appendix: The Dismantling of Nuclear Facilities.

Each chapter can be read independently, because the chapters are built around a central topic. The chapters share an overall coherence and there is a logical progression in the presentation of the themes: the methodological aspects of economic calculation are discussed in the first chapters, while the world perspectives of nuclear energy are analyzed in the last chapter of the book.

This book is intended for a wide audience with a minimum scientific knowledge and wishing to understand the details of the calculations of the various costs of nuclear energy, or to enrich their understanding of this problem: students, researchers, professionals in the energy sector (companies, ministries, communities, etc.) will find a large amount of information and be able to form an opinion regarding the economic aspects of nuclear energy.

The intention behind this work is to keep it as academic as possible, methodically describing the hypotheses and uncertainties that relate to knowledge or theories of the future, as well as to technical or political choices. The profile of the authors is heterogeneous, striving to summon a wide spectrum of specialists, from university teachers to pilots of major industrial projects.

Readers will undoubtedly appreciate the nuances in the approach to given topics, and while some "opinions" may permeate certain introductory sections, the entire work constitutes an unprecedented scientific resource. This book aims to provide factual, accurate, comprehensive and accessible elements, and contribute to an informed and dispassionate debate meeting current energy and climate challenges.

April 2023

[illegible]

Each chapter can be read independently, because the chapters are built around a [illegible]

1

The Evolution of Nuclear Energy in the World and in France

Daniel IRACANE[1], Stéphanie TILLEMENT[2] and Frédéric GARCIAS[3,4]

[1] *CEA, Saclay, France*

[2] *LEMNA, IMT Atlantique, Nantes, France*

[3] *IAE Lille, France*

[4] *Université de Lille, France*

1.1. Introduction

Nuclear energy has the advantage of being extremely concentrated, which makes it an attractive source of energy. However, it also requires harnessing the risks associated with such concentration.

For several decades, the equilibrium between the advantages and disadvantages of this technology has been the subject of bitter controversies in many countries. Safety, waste management and decommissioning are major issues whose long-term uncertainties fuel debates. These topics are at the heart of a know-how consolidated over the decades and reinforced by intense international exchanges, dictating the state-of-the-art in the field. This international state-of-the-art receives various nuances depending on the country and is translated by national legislation and regulations. Despite their critical importance, these topics fall under the broader question of the "nuclear know-how" and will not be addressed in this first chapter.

Nuclear Economy 1,
coordinated by Jacques PERCEBOIS and Nicolas THIOLLIÈRE.

The intention here is to evaluate the position of nuclear energy in the energetic-industrial policy of a certain number of countries, both in terms of motivation and management.

The consumption of considerable amounts of energy is the essential basis of our way of life and shows a perfect correlation with the world's GDP. The debates on energy policies are lively and choices even more difficult in a world which has to cope with climatic emergencies and which exceeded the conventional oil production peak in 2006. The tensions, not to mention the antagonism, between availability constraints, a low-carbon economy, strategic security and socially acceptable costs have never been so palpable. New energy technologies are tools which are all contributing to the evolution of electricity mixes.

But will nuclear energy remain a significant asset in the solutions toolbox deployed by our advanced societies in order to meet *both* the massive citizen demand for affordable energy and climate commitments?

International agencies (OECD, IAEA) advise on the need to use all the technologies at hand – including nuclear energy – since energy-related issues today seem almost insurmountable and terribly urgent. Despite this, many societal actors believe just as legitimately that our world should work because of new energy technologies and innovation, and dispense with controllable (fossil, nuclear) sources.

While innovation is the key to an unknown future, it is impossible to peremptorily demonstrate that a world based on renewable energies cannot meet the needs of our societies.

However, the deployment of the technologies required for collecting wind power and solar energy, whose nature is diffuse and intermittent, raises new challenges when such deployments take place on a very large scale. A country like France consumes an average electrical power in the range of 50–60 GW. Estimates agree to show that a 100% renewable system, based on the production of hydrogen to compensate for intermittency, would require the deployment of around 300 GW of technological objects (wind turbines, solar panels, electrolyzers and fuel cells), namely, five to six times the energy needed. While wind power and solar energy are renewable, the technology to extract and concentrate them is not. This poses a major challenge and systemic uncertainties, for example, in terms of the materials needed. It is therefore just as premature to guarantee that we can do without a carbon-free, concentrated and controllable energy such as nuclear power.

Confronted with a societal debate of primary importance, it is crucial to examine the trajectory of a few emblematic countries in their willingness and their ability to include nuclear energy in their technological and energy policies.

This first chapter will describe the evolution of nuclear energy, first at a global scale, before focusing on the French case.

1.2. Nuclear power in the world

In its pioneering phase, nuclear science and technology shared a parallel development within a few countries. At first, it was necessary to overcome the scientific challenges resulting from the discovery of the neutron in 1932, and fission in 1938. In these countries, the Second World War marked the starting point of an often dual technological adventure, which materialized in the 1960s through the proliferation of national programs widely exploring the different solutions for producing electricity. The 1974 oil crisis led to a new paradigm shift associated with the supply security challenges, leading both to the choice of the best technologies and their rapid deployment. All of the countries having followed this path share the same development model built around a State determination and framework, concomitantly encouraging the emergence of industries and national regulatory systems. The last two decades of the 20th century were marked by the stabilization of the world's nuclear fleet after a period of strong growth, and above all, by the liberalization of the electricity markets (Deffrennes and Iracane 2020).

Following the homogeneous heritage described in section 1.1, the trajectories of the different countries equipped with nuclear energy reactors has diverged since the 2000s. This is not only related to the public perception of nuclear energy and the economic and legislative framework given to the electricity market, but also due to varying geostrategic visions associated with the key domain of energy. More recently, climate issues, the breakthrough of shale gas, as well as the sustained growth of intermittent solar energy and wind power, have further increased the complexity of the energy subject in all countries.

It is therefore interesting to examine the different management and development modes of nuclear technologies around the world. The perception of these different dynamics offers a salutary perspective on a subject often caught in frozen debates.

The following paragraphs will provide examples of various management and development modes – to a certain extent each embodied by a particular country – which will then be considered as an illustrative basis. The illustration by country in no way suggests that the country in question favors a monolithic approach to nuclear power. On the contrary, to varying degrees and more or less explicitly, all the management methods illustrated below affect each country, which invites apprehending the nuclear subject from a natural multiplicity of viewpoints.

1.2.1. *Creating long-term visibility seducing the market to invest: the case of the United Kingdom*

Nuclear technology is capital-intensive and its associated investments are only possible within the framework of a stable long-term vision. The major historic nuclear programs, as well as the more recent one in China, were the result of a State readiness embodying this long-term vision instead of focusing on a market dynamic. In contrast, the United Kingdom implements an emblematic approach according to which the State's role lies in establishing the relevant conditions so that the market can operate until a new generation of nuclear reactors is deployed.

The United Kingdom's nuclear trajectory is characterized by large-scale transitions between a pioneering period in the field marked by British leadership, a second phase of brutal abandonment, and finally, the current juncture signaling a voluntarist and politically assumed reconstruction.

With the depletion of its hydrocarbon reserves in the North Sea and the aging of its energy infrastructure, the United Kingdom turned to low-carbon, nuclear and renewable energies, the fruit of a cross-party consensus. In 30 years, the electricity mix massively shifted from carbon to gas supplemented by wind power and solar energy, the nuclear component remaining globally stable. Although in 1999 the United Kingdom surpassed its gas production record, in 2002 its production began to fall.

After a long period of hesitation, striving to find an energy solution primarily based on renewable energies, successive governments built a political project integrating the renewable component and the nuclear component into the same "green economy", motivated by three simultaneous commitments: reducing CO_2 emissions by 2050, supply security and cost reduction for the consumer. In September 2016, the signing of the *Contract for Difference* for the Hinkley Point C project involving the construction of EPRs was the concrete kick-off signal for British nuclear renaissance.

The compatibility between the deregulation of the electricity sector and electronuclear technology continues to be a key issue in the use of this technology. The United Kingdom is a textbook case in this respect. The first announcements regarding the privatization of the nuclear sector date from 1988. In 1995, the

"Review of the future prospects for nuclear energy" report in the United Kingdom (Danby 1996) definitively concluded on the interest of a transfer to the private sector for the industry, the consumer and the taxpayer. This was followed by the creation of British Energy in 1995 and its privatization in 1998, operating eight nuclear energy plants. Later, British Energy was destabilized by the priority given to shareholder compensation over investment and the company's international adventures. The new market structure (New Electricity Trading Arrangements [NETA]), launched by the government in 2001 with the intention of promoting competition, caused wholesale electricity prices to fall and sent British Energy into the red as early as 2002. Following unsuccessful rescue operations, the government withdrew from the company in 2006, which was bought by EDF Energy in 2008. In 2020, with eight nuclear energy plants, 35 wind farms including two off-shore facilities, one gas-fired and two carbon power stations, EDF Energy became Great Britain's leading electricity supplier, meeting 20% of the country's needs.

> The United Kingdom was a pioneer country in nuclear energy with two experimental piles commissioned in 1947 and 1948, and a first Advanced Gas-cooled Reactor (AGR) coupled to the network in 1963. As in other countries, including France, the debate for the choice of future power reactors between national technology and American technologies was fierce. In 1965, the British government chose the national AGR sector. But in 1979, the Thatcher administration endorsed a policy shift, authorizing the construction of a pressurized water reactor at Sizewell, using American technology. This was the last reactor built in the United Kingdom until the official start of construction work on two EPRs (European Pressurized Reactors) by EDF Energy and the Chinese enterprise CGN at Hinkley Point C, in September 2016.
>
> In 2006, Tony Blair's administration announced its intention to launch a program for building new reactors in the United Kingdom, aimed at replacing the existing stations and increasing the share of nuclear energy in the electricity mix, in order to reduce British greenhouse gas emissions. Several projects were proposed, planning for the construction of different types of reactors: the French EPR, the Japanese-American AP1000 and ABWR, and the Chinese Hualong. The paths followed by France and the United Kingdom in terms of nuclear energy share many similarities. The choice of the technological sector for a massive deployment in the 1970s reveals a key bifurcation; the choice of the British for their national low-power density reactor technology largely accounts for the setbacks the industry experienced during its privatization.

Box 1.1. *Technical history of nuclear energy development in the United Kingdom*

Together with France, the United Kingdom is one of the few Western countries to have deployed industrial leadership in the field of the fuel cycle. British Nuclear

Fuels Ltd[1] (BNFL), the flagship of the English nuclear industry, exploited the benefits of national public investment and became a world leader, then quickly lost momentum, until its complete dissolution. Following its transformation into a public limited liability company in 1984, BNFL became one of the largest nuclear companies worldwide, providing its services at all levels of the fuel cycle, acquiring Westinghouse Electric Company in 1998, and then the nuclear portion of ABB (ASEA Brown Boveri, a Swiss-Swedish company). As from 2005, BNFL became gradually weakened by its dismemberment into several national organizations and by the sale of its assets (sale of Westinghouse in 2006 to Toshiba), until being completely dissolved in 2010.

One may observe that after a first-rate pioneering phase, the privatization of an industrial sector seeking to access a classical market was a flagrant failure.

The "Business, Energy and Industrial Strategy" ministry created in 2016 signified a remarkable turning point, with a renewed desire to develop an industrial strategy. In the nuclear field, this desire materialized in 2018 via the Nuclear Sector Deal (HM Government 2018), opening up prospects in the field of human resources and skills management, infrastructure deployment and innovation.

In view of meeting the challenge of the massive renewal of its nuclear fleet between 2020 and 2030 and honoring its commitment to reduce greenhouse gases (carbon neutrality by 2050), the British government is supporting the relaunching of a program for the construction of nuclear energy plants and striving to create favorable conditions for this program.

In the field of safety, in order to accommodate a broad technological offer, the government instructed the regulators – the Office for Nuclear Regulation (ONR) and the Environment Agency – to create a Generic Design Assessment worksheet, in view of facilitating the regulatory visibility of emerging commercial proposals. The role of the safety authorities is fundamental to enable the emergence and penetration of new technologies. Given their independence, a clear political framework is necessary for the safety authorities to set up a transparent process when working with technology suppliers upstream of deployment decisions. Equivalent approaches motivated by the desire to promote national innovation or to accept innovative foreign technologies are observable in the United States, as in Canada.

1 British Nuclear Fuel Ltd (BNFL) was established in 1971 to manage power stations (Calder Hall and Chapelcross), the Magnox fuel reprocessing plant (Windscale, now Sellafield), gaseous diffusion enrichment (Capenhurst) and fuel fabrication (Springfields).

In terms of financial protection, the British government created conditions for long-term visibility, enabling investors to secure their business plan. With the Contract for Difference, in 2013 the government introduced a strike price of £92.5/MWh (indexed to inflation) over a period of 35 years from the start of operations. For another EPR construction project (Sizewell), a new financial architecture (the Regulated Asset Base or RAB) is being studied, whose aim is to reduce the price per kWh. This mechanism, quite innovative for nuclear energy, enables an independent regulator to periodically review the situation and set prices.

The Hinkley Point C project marked a second emblematic change. In 2016, the Chinese CGN electricity supplier acquired a 33.5% stake in the project, alongside EDF, thus marking the first significant Chinese step into the Western nuclear field. This paved the way for another project by CGN with the support of EDF involving the certification and construction of a 100% Chinese design reactor (HPR 1000 or Hualong) on Bradwell B site.

Great Britain's rise to power in the nuclear field is being catapulted by its plans to build Generation III nuclear reactors, while relying on a market with a facilitative public framework. This framework also implies a proactive long-term support policy for the sector, with priority given to human resources management and actions in favor of open innovation. A continuous production of initiatives can be observed at government level. Their goal is to rebuild a global nuclear system: the creation of a support structure for the definition of government policies, such as the diligent Nuclear Innovation Research Office (NIRO), the production of a "Nuclear Innovation Program", the launch of the "Industrial Strategy Nuclear Sector Deal" in 2018, the call for projects for the development of advanced modular reactors (AMR).

After the disengagement of the 1980s and 1990s, the starting point for this reconstruction is undoubtedly low, and it will take long for this policy to materialize a new British leadership in the field. Government documents, such as the Energy White Paper (HM Government 2020a) and the Ten Point Plan for a Green Industrial Revolution (HM Government 2020b), released at the end of 2020, clearly state the British goals in terms of decarbonization and consistently classify nuclear power alongside offshore wind turbines as clean energy. This contrasts with many other countries, in Europe and elsewhere, whose technical capacities have remained at a high level in terms of means, but whose dynamism is losing momentum in the context of hesitant national policies on nuclear energy, lagging behind in terms of innovation and maintaining competencies.

1.2.2. *Nuclear activities at the heart of State integration and at the service of economic diplomacy: the case of Russia*

Unlike the British case, Russia embodies the historic nuclear paradigm fully integrating all the functions within the State corporation, Rosatom: not only research, engineering, manufacturing, production and management of nuclear materials from mine to fuel, electricity supply, but also institutional prerogatives at a quasi-ministerial level. After being a competitor among others by the end of the first decade of the century, 10 years later Russia became the undisputed leader in the field, confirming the performance of an integrated state-owned enterprise (SOE) for an activity such as nuclear power.

The State corporation, Rosatom, created in 2007, brings together more than 300 companies and employs 25,0000 people. Rosatom's general management has direct control over the entire nuclear industrial sector, both within the national economy and on export markets.

This is a powerful State monopoly which benefits from a voluntarist commitment on the part of the government, not only motivated by financial resources for export, but also by the availability of a tool strengthening its economic and geopolitical diplomacy in the energy field.

It is interesting to call into question this organizational structure based both on the merger of technical functions as well as on such a high level of politicotechnical integration. At first glance, this scheme may seem reminiscent of a bygone era and only few similar examples can be found in other fields and other countries.

The observation cannot be contested though. In barely a few years, Rosatom has become the undisputed world leader in the field, while its main competitors have withered away.

Yet, during the 2000–2010 decade, the games were far from over. That era corresponded to the so-called "nuclear renaissance", characterized by significant growth prospects in the nuclear fleet around the world, not only in large countries such as China and India, but also in many emerging countries, increasingly aware that the availability of an efficient source of energy could enable them to climb the steps of development.

In 2001, France merged several companies into the Areva group which could boast an enviable order book by the end of 2008, 71,000 employees (55% of whom were abroad), and planned to recruit 2,500 executives in 2009 alone.

Other Japanese and American powerful players, such as Toshiba/Westinghouse, were starting to enter onto the international market. With the support of the United States, the South Korean KEPCO (Korea Electric Power Company) made a sensational entry into the international market. Its major achievement was signing the sales contract for four reactors to the United Arab Emirates at the end of 2009.

Which could be the underlying reasons for the disappearance of the Areva group in 2018, and the loss of the domineering positions by major American, Japanese and Korean players?

The size and monopolistic character of Rosatom do not seem to be a decisive reason because all the industrialists in the field naturally create consortia providing all the resources and competencies required for major nuclear projects. While it is true that the technical products of competitors are different, they all belong to the same technological generation. The setbacks of the American and French construction projects appear more as iconic scenarios rather than causal factors. An array of factual elements led to the weakening of Rosatom's competitors. Suffice to mention, for example, the Fukushima Daiichi nuclear accident, which marked a political turning point, and the Basel III agreements, which reformed the banking sector after the 2008 financial crisis and made it more difficult to finance major nuclear projects in the Western world.

Beyond these factual elements, a root cause of the divergence of trajectory between Russia and its competitors in the nuclear field probably lies in the intertwining between the technical and political spheres, which is clearly more powerful in the Russian model:

– The value of nuclear projects is long-term, offering great stability to the electricity mix, but also requiring stability in political and financial conditions. In the United States, the power plants which close down due to economic reasons are set in deregulated market areas, revealing that the market alone is insufficient when it comes to supporting a field of activity such as nuclear energy. The long-term vision of States and the transpartisan stability of this initiative are structuring elements for a viable nuclear industry.

– When this vision is promising, the States undertake national programs, making visible and deterministic prospects and domestic investments in the nuclear field. The existence of a long-term perspective on the domestic market is a cornerstone for the industry, because of which the entire supply chain can invest in skills and innovation. This has direct consequences on the performance of national construction sites and export credibility.

– The financial aspects are determinant for such a capital-intensive industry. The reinforced regulation of the financial field is in itself a source of complexity, as it involves the capping of public loans and the coverage of these loans by State guarantees. Regardless of any technical differences between countries as to the implementation of the international regulatory framework, this is a clear indicator of the greater or lesser determination of States to engage in nuclear export. This State's readiness makes all the difference in the negotiations and the risks involved in decision-making.

The above-mentioned elements prove that, given its tropism, the nuclear business sector will have to fiercely struggle to achieve the performances required for its survival, unless it is backed up by a long-term vision and a promising State policy. This explains why the United States and Europe have not been able to maintain their leadership.

In contrast to its competitor countries, Russia has maintained a stable, strong and long-term vision for its nuclear activity, encouraging its industry to thrive in a clear strategic context. This includes the definition of a national market structured by the replacement of 10 of the 37 existing light water reactors, providing great visibility to the entire Russian industrial supply chain.

Russia has also engaged in the development of small modular reactors (SMRs), already used for the icebreaker fleet and which can also be installed on land or on a barge (55 MWe RITM-200 and 120 MWe RITM-400 reactors).

At the same time, Russia maintains its closed cycle strategy and has become an undisputed leader on the subject of fast neutron reactors, whose know-how is attested by the BN-600 operating reactors, the BREST-300 units under construction and the planned BN-1200 reactors. Furthermore, France and Japan, which shared this leadership, no longer have facilities in the area and have stopped their construction projects. The United States is relaunching a project to build a test reactor with the aim of redeveloping competencies in the field, without however having defined a strategic vision on the use of this technology; only the future will show whether a project of such magnitude can materialize without an encompassing vision.

On the other hand, relying on the strength of its national dynamic, Rosatom exhibits credibility and dynamism in exports, which are evidenced by an order book of 36 reactors for export in 12 countries and a value of $133 billion in exports scheduled for the upcoming 10 years. Rosatom's export projects imply economic and geopolitical challenges because they establish long-term relational bridges with partner countries, thus completing the Russian geopolitical portfolio in the energy sector. In this perspective, Rosatom benefits from strong diplomatic and financial support at the highest level of the State to materialize its export offer. At the same time, Russia can grant loans covering the greater portion of operational costs or even engage in direct investment (e.g. projects managed under the Build-Own-Operate modality by Hanhikivi in Finland and Akkuyu in Turkey). In the same spirit, since 2014 the State has mandated and allocated resources to the safety authority Rostekhnadzor to provide assistance to countries acquiring Russian nuclear technologies, which is a key factor in attracting the interest of countries wishing to access nuclear energy.

Russia's inevitable challenger is of course China. The Russian trajectory in China is parallel to that of France and other Western technology providers. Russia has deployed several reactors in China (4 VVER-1000 in Tianwan) and signed the delivery of four others in 2018. Rosatom's role is gradually being shadowed by the rise in competence of the Chinese ecosystems and by technology sinicization, which will make the sales of other reactors highly unlikely. This will be followed by the rise of China in the export of national technologies, probably on the basis of head-on competition with Russia, as the two countries offer similar arguments: mastery of technology thanks to benchmarking, the ability to advance funding, State support, etc. However, it is difficult to anticipate the balance of this future stage because the competition between the two countries will be part of a framework outstretching beyond the field of nuclear technologies.

1.2.3. *The concrete implications of political change: the case of South Korea*

After Russia, South Korea provides a mirror illustration of the structural link between nuclear energy and public policies in countries developing or using this technology.

As a matter of fact, South Korea made a sharp political shift in 2017 with the coming to power of the Liberal Democratic Party and the election of President Moon. Since this political change, the Korean State has carried out both a policy of gradual exit from nuclear energy in its electricity mix and a supporting policy for the export of its nuclear industry. It will be interesting to study this country over time to factually measure the impact of this duality on the nuclear industrial sector.

This duality is already limiting Korean export credibility by raising natural questions from potential customers who are looking for long-term reliability from their suppliers. The Emirati electricity supplier ENEC, operator of the Barakah reactors built by South Korea, publicly expressed this questioning. This might have influenced its decision not to award the exclusive long-term maintenance contract for the Barakah reactors to South Korean manufacturers, even though for Korean industry this contract represented a key element regarding the project's profitability. In this context, and despite the undeniable success of the Barakah project (four reactors with a 1,400 MW potency, among which the first became operational in mid-2020), as well as the technical and political support of the United States, South Korea has not won any export contracts since 2009, although its goal was to export six reactors before 2025.

Even if future options remain open, like other Western countries, the case of South Korea may illustrate that the absence of a national market and a coherent policy limits the viability of a nuclear industry capable of exporting major construction projects. Since 2017, almost half of the main companies in the field have announced a reduction in their workforce. The recruitment rate of new engineers has decreased accordingly. In a country where the access cost to university is very high, the number of students in the 18 universities with a nuclear engineering department decreased from 893 in 2018 to 457 in 2019.

The gradual abandonment of nuclear energy in South Korea is defined by the non-renovation of operating licenses (40 years), which will result in the closure of 12 reactors by 2030. As South Korea struggles to define and implement a management policy for its spent fuel, the closure of certain power stations could even be faster.

As in several countries in Western Europe, South Korea displays a dual energy agenda involving the abandonment of nuclear energy and carbon. Again, the Korean case will be interesting to follow. With a highly concentrated population in large metropolises (the urban area of Seoul has a population of 25 million inhabitants, that is, 40% of the country's population), a poorly favorable geography for the deployment of renewable energies, a quasi-island network and a powerful production industry, South Korea's energy equation is difficult. At the end of 2020, South Korea announced its carbon neutrality goal by 2050 and its ninth electricity program plan (covering the period until 2034). This plan clearly shows a massive transition to liquid natural gas (+50% with the replacement of 24 of the 30 carbon power stations shut down), having a significant impact on the cost of electricity and energy independence.

South Korea's economic boom was accompanied by cheap subsidized electricity for consumers. The population is little mobilized on energy savings and, more generally, on the subject of energy. The economic tension is high for the operator Kepco, tugged between a massive increase in production costs (transition to natural gas and renewable energy) and politically imposed low tariffs.

Regardless of how it is conducted, the transition to a more realistic pricing policy is raising greater societal and political awareness about the energy issue, which is the foundation for countries to build lucid and robust strategies.

1.2.4. *A new innovation paradigm at the service of reconquering the market: the case of the United States*

Western cradle of electronuclear technologies, the United States have long lost their technological leadership in the field. But, in the context of a growing technological cold war with China, it is difficult to imagine the United States accepting the Russian leadership in the field or the irrepressible Chinese dynamics, without a counter-reaction. The United States is then confronted with the question of the strategic path that will enable them to regain leadership in the international nuclear market.

Most of the 300 pressurized water reactors (PWRs), that is to say, two-thirds of the 443 reactors in operation worldwide in 2019, have been designed based on the concept developed by Westinghouse[2]. The end of the period as a national territory supplier led to a severe decline in American industry. This was illustrated by the sale of Westinghouse's nuclear assets in 1998 to the British company BNFL, resold to Toshiba in 2007, then in 2018 to a Canadian asset management company (Brookfield Asset Management Inc.).

In the recent period, nuclear energy has been confirmed by the American administration as a sovereignty technology, which has motivated an investment revival in nuclear technology. Now thrust into a challenging role, the United States relies on innovation to regain leadership in the field.

The global electronuclear industry has reached a high degree of maturity, with a cumulative experience over five decades equivalent to 18,000 reactors/year. The trajectory was marked by three major accidents[3], each signaling a major stage towards improving safety, both in technological and organizational terms. But this industrial and regulatory maturity disguises, or perhaps even induces, a major structural weakness in the field: the drastic slowdown in innovation and the renewal of technological generations.

2 Westinghouse designed pressurized water reactor technology by virtue of the compactness it offered, and equipped the first nuclear submarine, the Nautilus, in 1954.

3 These were the Three Mile Island (USA, 1979), Chernobyl (ex-USSR, 1986) and Fukushima (Japan, 2011) accidents.

Although the term "innovation" was not used, the pioneering decades of the 1960s and 1970s gave the United States the possibility of experimenting with a variety of electronuclear technology solutions. The cycle between the ideas and the manufacture of prototypical models at a large scale only lasted a few years. The transfer from research to large-scale industrial deployment took less than 10 years.

Development programs were implemented by large State agencies. The two personalities shown in Figure 1.1 were the leaders of NASA and the Atomic Energy Commission. These two agencies succeeded in creating a first-class industrial reality in the nuclear field in less than two decades.

Figure 1.1. *Atomic Energy Commission (AEC) Chairman Glenn Seaborg and NASA Administrator James Webb in July 1961*

Innovation is again of the utmost importance to optimize electricity production costs with permanent (or better) reliability, in order to provide solutions to technological obsolescence, promote a sustainable fuel cycle, accomplish waste management solutions, and, globally, ensure that nuclear technology meets societal needs.

The global deceleration in nuclear innovation is due to several concurrent reasons:

– As happens with all industries where risk control is a top priority, the entry ticket for any new technology is very costly because it involves raising the confidence level of old technologies. Technological innovation, as well as its accompanying regulatory framework, must be built coherently, which implies harmonizing a large ecosystem of stakeholders. While academic and industrial laboratories continuously deliver their share of ideas and solutions, penetration into an industrial system is lengthy and subject to extensive technology qualification periods (a decade or over).

– During the pioneering phase of nuclear power, the alignment between national priorities and the integration of scientific, industrial and political worlds was the pillar of efficiency, with consistent and swift decision-making processes. At present, however, the political hesitations specific to nuclear energy in many countries are weighing on market prospects and constraining the industry's ability to project itself into ambitious and long-term developments, limiting investments in research and development.

– In countries where innovation processes are losing their vitality, deceleration phenomena in technological development are reinforced by the erosion of research infrastructures and associated skills.

As a response, the United States increased public investment to revitalize the nuclear ecosystem:

– In the 1980s, the weakness of educational activities led to a reduction of over a factor of 2, both in public funding, in the number of nuclear engineering departments and in the number of students. Since the end of the 1990s, the US Congress have decided on a significant increase in funding for universities, resulting in a tripling of the number of students in nuclear science and technology in 10 years (OECD/NEA 2012).

– More recently, sustained investments have been granted to national research centers for the construction of infrastructures specifically open to university programs, subsidized up to €50 million and over, per research center annually.

This public investment in research and education was supplemented by a massive support for industrial investment in innovation. The Department of Energy (DOE), echoing its GAIN (Gateway for Accelerated Innovation in Nuclear) program, extensively supports the industry via the calls for projects in the field of innovative reactors or accident-resistant fuels. It is then up to the industry to choose disruptive paths, either by the major conventional nuclear manufacturers, by the major non-nuclear industrialists or by a whole group of start-ups. While the ecosystem's financial irrigation produces many concepts with no future, it also generates concepts that are now taking shape, such as the NuScale SMR project, which embodies the American entrepreneurial and innovative spirit in the nuclear field, as well as the boiling water BWRX-300 project, by GE Hitachi.

This is a major paradigm shift in the field of innovation, specific to the United States. In the space field, this paradigm is perfectly expressed by the technological revolution brought about by the SpaceX company in less than 20 years, with the support of NASA. In the historical model, public agencies developed the vision and directly piloted its implementation by means of extensive industrial subcontracting. The new model is based on an organization where public agencies define the main priorities and co-build a new technological ecosystem with a few industrial partners. It is difficult to say whether the success encountered in the space sector will become extensive to nuclear power, and even more uncertain, to interpret it as a model which can be generalized in Europe or elsewhere. But, to say the least, it represents an "innovation in the art of innovating".

Globally, DOE budgets devoted to nuclear energy experienced a significant increase (+85%), rising from $800 million in 2011 to $1,500 million in 2020. This increase in the government budget in support of nuclear energy materializes in the reactors field, as evidenced by a major commitment at the demonstration phase of new concepts, upstream of the investment cycle (in particular, targeting accident-resistant fuels), and finally on the infrastructures equipping the national laboratories.

At the same time, the electronuclear sector is under severe economic pressure. At the end of 2019, the United States had the largest fleet of electronuclear reactors: 96 units for a capacity of 98 GWe. Almost all the reactors have extended their lifespan to 60 years, with the launch of several units even reaching 80 years. The American nuclear fleet is now largely amortized, after having reached low average production costs of around $30/MWh in 2019. However, the competitiveness of this nuclear fleet is deteriorating under the economic pressure associated with shale gas, as well as the subsidies granted to solar and wind generators. In that sense, nine reactors have been shut down for economic reasons since 2013, and four others will be suppressed by 2025. This significant loss is considered detrimental and several States have extended the support mechanisms in vigor for intermittent energies to nuclear energy. At the same time, operators are optimizing the exploitation of their reactors, reaching record operating rates over 93% in 2019.

Innovation is being promoted both by industry and by the United States administration to support the operation's profitability. The development of new types of fuels, which are safer and can help simplify systems, is an emblematic example of an industrialist-driven program supported by the DOE. Innovative ideas abound and the public funds invested by the DOE accelerate the associated developments, as shown for example by the Advanced Reactor Development Program (ARDP).

For reactors in the last part of their lifecycle, these innovations should be deployed very quickly, so that they can influence the economic trajectory of reactors in time. Nevertheless, the maturation time for innovation in the nuclear field – from the initial idea to its penetration into the market – has become very long. This means, for example, that the development of a new fuel can take well over 10 years. As a result, the date with innovation and the current nuclear fleet is likely to be missed. The American administration and the safety authority have mobilized once again to tackle this problem. It is still too early to measure the results, but the subject is critical.

The qualification and safety demonstration for innovative solutions are a systemic problem for technologies with high safety requirements. The safety reference framework is the result of a long process based on largely consolidated experience feedback at an international scale. To benefit from an innovative solution, it will be necessary to establish a safety demonstration in the absence of experience feedback, which makes the authorization decision quite challenging. By way of illustration, note that the United States is the last major nuclear country not to have adopted digital command-control, which is nonetheless a mastered technology. It is necessary to work in parallel on technology innovation and on the evolution of the safety reference framework it implies. However, this contradicts the independence and required distance the safety authority should keep in relation to technology promoters. One solution would be to work on a safety reference framework at an international scale in order to protect the independence of national authorities. This difficult and limiting problem for nuclear power has been openly addressed in the United States and it will be very interesting to observe its evolution.

The American administration is mobilized to regain leadership in the design of new types of reactors. The motivation in this case seems essentially geopolitical. In the classic field of high-power pressurized water reactors, Russian and soon Chinese domination is too established to be challenged by the United States. However, there have been attempts by the DOE to put the AP1000 up for export (Poland, Czech Republic), despite the delay incurred by the Vogtle construction site and the doubling of its cost (estimated at $28 billion in 2019 for two AP1000 reactors).

The solution for the United States lies in radical innovation with new reactor concepts for new markets and applications, which do not necessarily involve power generation. Thus, the United States has created an ecosystem exploring a large number of reactor technologies in parallel in order to respond to the potential new market segment of SMRs.

This bet is based on reducing the reactor's unit power (ranging between 50 MWe and 300 MWe) in order to guarantee safety more simply, as well as to facilitate financial engineering, which is currently considered as the main barrier to nuclear entry. These two advantages are counterbalanced by a potential increase in the costs of the service rendered ($/kWh), because the costs of civil engineering and major components decrease less quickly than power. This is what justified the strong powers paradigm, including the French EPR, with its 1,650 MWe of power. To find an attractive cost, the SMR concept invokes the prospect of mass production in dedicated factories; this manufacturing innovation could be a game-changer and a major asset in nuclear energy.

This nonetheless raises the question as to the implementation of a heavy infrastructure. This could only be motivated by a market of several dozen SMRs, which would only be economically attractive if the infrastructure already existed. A world with many small and medium power reactors, mass-produced at especially dedicated manufacturing facilities, is perfectly conceivable. The ability of the market to establish a viable path for this universe of SMR reactors is still unknown to this day.

Currently, SMRs are considered a solution for the future and the United States holds a leading position in this potential market, with concepts such as NuScale (whose architecture is innovative, but relies on traditional technologies), and many other technologically disruptive notions that are being explored by start-ups.

1.2.5. The severe nuclear accident or the possibility of a sudden halt to the use of nuclear energy: the case of Japan

Given its powerful economy and a total lack of its own resources, Japan has embraced a long-term nuclear vision since the 1950s. The tragedy of the 2011 earthquake off the Pacific coast of Tohoku abruptly shattered national cohesion on the subject. Even though the justification of a nuclear energy program remains significant and is supported by the government, the future of the Japanese nuclear program is yet to be decided.

Japan's recent trajectory is that of a leading power lacking natural resources and whose exporting economy heavily depends on large-scale fossil fuel imports. Therefore, energy independence has long been a priority for Japan, especially considering that the greed of its Chinese neighbor for natural resources reinforces its restlessness.

Nuclear energy appeared in Japan in 1955 in response to this context, and achieved a first milestone in 1966, when its first reactor was connected to the network. Later, it was the oil shocks of the 1970s that pushed Japan down the path of massive nuclear industrialization with three major manufacturers: Mitsubishi Heavy Industries (MHI) associated with Westinghouse, Toshiba and Hitachi, both associated with General Electric.

These three manufacturers, powerful competitors in Japan, have developed alliances, not only to build the Japanese nuclear fleet, but also to strengthen their international presence. This is how Hitachi joined forces with General Electrics within the GE-H consortium to promote boiling water reactor technology (ABWR, ESBWR). For its part, in 2001 MHI concluded an agreement for the co-development of the American AP1000 reactor by Westinghouse, which was later abandoned in 2007 when it lost the contract to buy the American industrialist at the hands of Toshiba. Since then, MHI have entered into an alliance with the French Areva through the creation of the ATMEA consortium in 2007, whose intention was to design a Franco-Japanese 1,000 MW reactor. In 2018, following Areva's corporate restructuring, MHI became a shareholder of Orano with a share of 5% and Framatome for a value of 19.5%.

Thanks to its three reference manufacturers, Japan rapidly deployed 54 generating reactors which supplied 30% of its electricity consumption by 2010. In the same spirit of close cooperation with France, Japan has continuously invested in fast neutron and closed cycle technologies making it possible to improve the efficiency of natural uranium by factors of 50 and over.

Backed by a strategic need for energy, reinforced by climate commitments, with a large industrial sector motivated by export prospects, the Japanese nuclear energy program was part of a long-term ambition until 2010.

This trajectory was suddenly interrupted by the accident at the Fukushima Daiichi power station in 2011.

The 2011 tsunami killed nearly 20,000 people. According to UNSCEAR[4], the radioactive fallout from Fukushima's nuclear accident has not caused any casualties

4 UNSCEAR is the "United Nations Scientific Committee for the Study of Ionizing Radiation", a United Nations organization created in 1955, which today brings together 27 member countries, regardless of whether they use nuclear technology or not. UNSCEAR publishes authoritative reports under the control of the UN General Assembly.

to date, be that among workers or civilians (UNSCEAR 2020). On the other hand, it involved the displacement of over 150,000 people, creating many dramatic situations for the evacuees, including deaths related to emergency management, with a major and persisting psychological, material and financial impact for the region.

The first of its kind, the nuclear accident was the result of an external cause and affected several reactors on the same site. It led to intense international cooperation among the safety authorities around the world, which quickly drew lessons from the Japanese accident. In the months following the event, the safety framework for all nuclear facilities around the world was reassessed in relation to external hazards. Second, adjustment plans were implemented so as to strengthen resistance to external attacks: bunkering of emergency diesels, ultimate heat sinks, improved crisis centers, reinforcement of reactor containment systems and protection of spent fuel stored in pools improved mitigation of the hydrogen risk, optimization of regional and national response capacities to crisis-ridden sites, etc. At the same time, the question of the interaction between nuclear safety culture and national culture has developed in international circles, the only ones capable of helping countries to realistically evaluate this important point.

The Fukushima Daiichi nuclear accident marked a new level in the safety of nuclear energy plants. International collaboration has contributed effectively to this analytical process and helped in the construction of a new reference framework.

The immediate impact of the accident on the energy policies of countries using nuclear energy in their electricity mix is relative.

In the days following the accident, Germany announced its final exit from nuclear energy, but it only consolidated and accelerated decisions which had previously been made abiding by the 2002 atomic law. The hesitations about the merits of discontinuing nuclear energy which had emerged in 2010 were swept away by public emotion, and the 2011 atomic law ratified the political decision regarding the nuclear power phase-out.

With the exception of Germany and Taiwan, countries featuring a nuclear program have maintained their trajectory with vast investment, in order to integrate the improvements learned from the accident's lessons. It is clear that the accident strongly influenced the perception of populations on nuclear energy and its long-term impact on the nuclear trajectory of countries remains an open question.

The decade 2000–2010 was known as that of "nuclear renaissance", with many emerging countries perceiving the access to nuclear energy as a development accelerator. This "renaissance" whetted the appetites of technology-supplying countries, which mobilized their diplomacy and industrialists in this direction. The Fukushima Daiichi nuclear accident clearly signified the end of this rebirth period. But an indicator does not necessarily amount to a cause. The barriers to entry hindering the access to electronuclear technology are not negligible: the need for long-term political stability, difficult financial arrangements to satisfy a capital-intensive technology, numerous legislative and technical prerequisites, competition with increasingly accessible natural gas, etc. In many emerging countries, these difficulties may justify the delays in decision-making and the loss of interest in nuclear technology, regardless of the 2011 accident.

Last, but not least, the case of Japan remains torn between a strategic imperative for energy independence and the trauma of the accident.

Japan's geography and its urban concentration limit its capacity for deploying wind power and photovoltaic energy. Its need for concentrated, controllable and carbon-free energy justifies its appetite for nuclear energy, even in the post-accident context.

Confronted with an unchanged strategic need, the government has confirmed its desire to use nuclear energy in the long-run by affirming that the share of nuclear energy by 2030 will equal 20%–22% (compared to 30% in 2010), and by supporting public and industrial research and innovation.

But contrary opinions are strong.

Ten years after the accident, more than 50% of the population wishes to discontinue nuclear energy. While in theory the green light from the safety authority suffices to have a reactor restarted, in practice, the consent of the local authorities hosting the facility is required. Of the 54 reactors in operation in 2010, 21 were permanently shut down and, of the 33 operational ones, only nine were actually restarted in 2020. In 2019, nuclear energy only contributed 7.5% to the electricity mix.

After Europe, Japan has entered a deregulation phase of electricity markets, which acts as an unfavorable scenario for capital-intensive technologies, since investors can no longer harness long-term market risks. This change in the economic paradigm is perhaps a more determining factor than the political uncertainties in the field and could motivate industrialists to lower their level of ambition and commitment in the nuclear field. Note that Toshiba has sold Westinghouse, Hitachi is seeking to resell its Horizon project in the United Kingdom and MHI has abandoned its ATMEA medium-power reactor project. Unsurprisingly, the future of nuclear power in a Japan dramatically marked by the Fukushima Daiichi accident is uncertain. This uncertainty in the nuclear field concerns its energy policy, its industrial capacity as well as its dismantlement policy, penalized by Japan's challenge to reduce its plutonium inventory.

The three major nuclear accidents marked significant steps in the establishment of a rigorous safety reference framework. They have also definitively shaped the mental representation of nuclear power in most of the world's population. It is arduous to determine the impact of these accidents on the nuclear trajectory of countries featuring this technology in their electricity mix. The reason might be that the use of nuclear technology is confronted with the difficult equation of energy systems having to trade off an excessive thirst for energy on the part of modern societies, the need to decarbonize this energy and a reasonable price. This equation offers little leeway and requires acknowledging all existing technologies, even though the detriment associated with each of them is fully known.

1.2.6. *China: the emergence of an international leader*

The landscape described above concerns the historical leaders of nuclear energy, which, despite having different trajectories, all originated during the pioneering era of this technology, over 50 years ago. In 30 years, China multiplied its electricity production by 10. The share of electricity production from carbon was multiplied by 1.75 between 2008 and 2018, reaching 10 times the whole of French production. With such a dynamic, foreign companies shifted their perception of China: at first, it had the status of a target customer, it then became a partner and finally, a formidable competitor. Nuclear power is no exception to this accelerated (15-year) structuring transition of the world market.

In the field of nuclear technologies, as in most economic sectors, the currently position adopted by China is a structuring factor in the global trajectory of technologies. Having metamorphosed from an emerging power – rich in opportunities and attracting global industry – to a leader in technological production and more recently in development and innovation, China can now place itself at the heart of the alliances promoting its own technological and political influence in the race against the United States.

In the field of electrical energy, the situation in China is the dominant term in the global electricity mix. In 2018, China accounted for 47% of the carbon consumed in the world for the production of electricity. The growth of this production has never been as strong as in the 2010s with a slope of additional 193 TWh every year, the equivalent of the commissioning of an additional 1 GW carbon power plant fortnightly. In 2019, carbon represented 58% of China's energy mix and 65% of the electricity mix. In 2019, out of global emissions of 33 Gt of CO_2, the production of carbon power plants amounted to 10.1 GT of CO_2 in the world, and 4.6 Gt of CO_2 for China. China is taking proactive action to reduce pollution, particularly in the cities, but modifying the energy structure of its productive system is still a challenge, as shown in Figure 1.2. The Chinese government has set an ambitious goal by announcing it will strive to reach the carbon peak by 2030 and carbon neutrality by 2060.

As in other areas, China invested in nuclear energy with the goal of testing and subsequently appropriating industrial know-how. This notably included French, American, Canadian (Candu) and Russian (VVER) technologies. The Chinese nuclear program is an industrial success. It is the result of a rapid increase in competencies with an ambitious construction program. Since 2005, it has focused on the construction of nearly 50 reactors.

The process of nuclear appropriation is well illustrated by Franco-Chinese cooperation, which began in the early 1980s with the Daya Bay and Quinshan reactors, and continued until the signing of the contract for two ERPs in 2007. After a learning phase of 8 years for the Daya Bay project until its commissioning in 1994, the sinicization of technology could start with an increasing localization of the supply chain and a growing leadership in project management, with the construction of Ling Ao, which entered into service in 2002. In a context of international technological diversification, in 2000 China signed a partnership agreement with France to set up a more balanced cooperation and experience sharing between EDF and CGN operators, marking the end of the phase of Chinese appropriation and a partnership based on an equal basis. China's double-digit growth in the 2000s multiplied its energy needs and accelerated its nuclear program. Reactors inspired in Daya Bay but substantially modified were quickly deployed with more than 70% Chinese components.

In the second half of the 2000s, while Europe and the United States had not built reactors for 20 years, China embarked on Generation III technology, with the purchase of four Westinghouse AP1000s and two French EPRs. In 2018 and 2019, China was the first country to connect two EPRs to a network, with a controlled delay and ahead of their European equivalents, whose development had started earlier. This success marked a new stage where China demonstrated its undisputed leadership in manufacturing and construction, an unsurprising result given the dynamics of its program.

Strengthened by the experience acquired with France and overcoming the difficulties encountered with the AP1000 and EPR projects, China developed the national Hualong design. China proved its full technological mastery in the field of reactors by launching on September 4, 2020 the fuel load for its first Hualong-1 reactor, a 100% Chinese reactor from its design to its commissioning. In 2019, the decision to build six new reactors was intended to inaugurate a new decade of reactor construction at a rate of six to eight reactors per year.

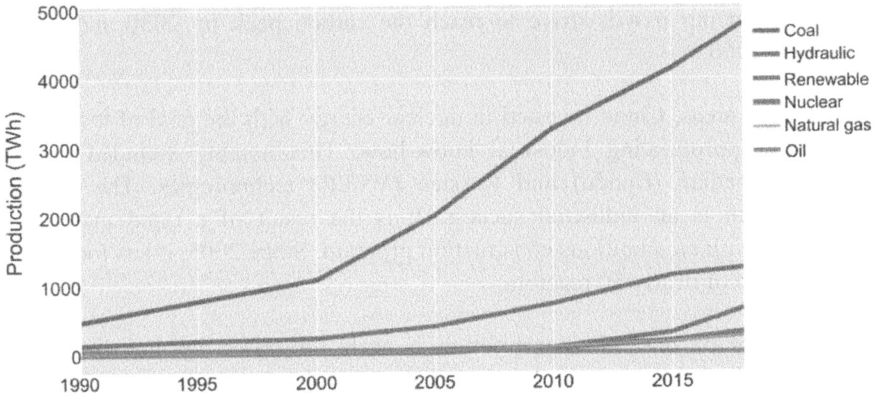

Figure 1.2. *Evolution of China's electricity production between 1990 and 2018 (source: AIE). For a color version of this figure, see www.iste.co.uk/percebois/nuclear1.zip*

China now naturally turned to export markets, both to serve its political influence and to supply its industry infrastructure. While its technological mastery and financial power are major assets, they are counterbalanced by political pressure from the United States over their allies and China's political image. In Western countries, China's export performance is limited and is reflected in the more captive market of its historic allies.

Maintaining the best level of the Chinese nuclear industrial performance would require launching a vast domestic program of new reactors, in coherence with the new double cycle Chinese policy which involves "running the economic engine by consumption and domestic projects", and not exclusively by exports, which are more difficult to control.

Like other economic sectors, nuclear power tends to be polarized by the Sino-American duality. The tension with the United States leads China strengthen the resilience of its nuclear industry in order to no longer depend on foreign technologies or suppliers. This is why certain domestic tenders are now closed to foreign suppliers.

China now has full control of nuclear energy generation technologies. The carbon neutrality goal by 2060 announced by China at the United Nations General Assembly in September 2020 will require a major shift in its energy mix. Such a change will imply not only the use of all technologies (nuclear and renewable ones), but also an acceleration in the deployment of these technologies. The press is making reference to about 100–150 GWe by 2035, that is to say, four to eight new reactors per year, mainly involving Chinese technology (Hualong type).

While China has succeeded in mastering the necessary know-how in the areas of reactors and fuel fabrication, the same cannot be said downstream of the cycle, where its investment is accelerating. China thus seems to have confirmed its choice for recycling the nuclear materials present in its spent nuclear fuels, initially in pressurized water reactors (thanks to the use of MOX fuels), and subsequently in fast reactors. To achieve this, China is investing on domestic technologies and, in parallel, on the negotiation of the sale of a used nuclear fuel processing-recycling plant by France.

Driven by the rationality of its State centralism and its existential need for clean electricity, China is not as hesitant as many other countries as to the use of nuclear energy. However, public acceptance becomes a key factor in the site choice for the deployment of nuclear infrastructure. In the field of waste management, China has several surface storage centers and a geological storage center. This is also the case for the installation of reactors on the river banks. Although the project remains at an impasse to this day, these sites respond to a significant local energy need.

As in other technological sectors, the trajectory of China regarding nuclear power followed a fast track toward technological mastery. It quickly tackled innovation, belying predictions of a saturation of its progress in the low tech, or even a simple manufacture-reproduction based on foreign recipes. By having created favorable conditions for the appropriation of nuclear know-how, China demonstrated its ability to support a complex, integrated and autonomous system.

In the nuclear sector, the differential between China and the historical leaders in the field has been largely reduced. Competition is partly determined by the ability to innovate.

In recent times, China has implemented several policies that contribute to sustaining its capacity for innovation. In the nuclear field, strengthened by the considerable resources this country can mobilize, there is a systematic approach to research encompassing all the reactor technologies currently under discussion in the world. The parallelism between the effort (number and diversity of sectors studied in parallel) and the intensity of such effort (until the implementation of operational prototypes) contrasts with the rest of the planet.

An example is the Institute of Nuclear and New Energy Technology (INET) at the University of Tsinghua, which developed the high-temperature reactor technology up to the construction of a 10 MWth experimental reactor (HTR-10 commissioned in 2003), and which pursued the program by designing and building two 250 MWth reactors (HTR-PM) with industrial partners, which were in the test phase in 2020.

The scale of financial power associated with political determination leads to implementing the breeding ground for tomorrow's technologies.

Nevertheless, innovation also requires individuals who are able to think about discontinuities. Limitations are applicable to China, not only in cultural terms with an education system that traditionally does not facilitate question-settling and intellectual inquiry, but also due to a constrained political framework which does not encourage openness. To compensate for this, China aims to attract part of the international elite, in particular by investing in the most advanced experimental and digital infrastructures. In the same spirit, China has welcomed projected structures of the world's best schools and universities to diversify its teaching models. It is still too early to measure the impact of such initiatives; they illustrate the awareness of public authorities on the subject.

While innovation requires an ability to think about breakthroughs, it also requires the ability to quickly transform an idea into the deployment of industrial products. On this point, China has an uncontested leadership by size effect and by its political system. It is very likely that China will drive all emerging nuclear technologies to the highest level, with the only limit being global competition in standards and areas of influence.

Unsurprisingly, the analysis of Chinese nuclear energy quickly refers to cross-cutting considerations in all technological fields. Even with few achievements in the international market, China has already acquired all the elements enabling it to provide leading leadership in electronuclear technologies. Its exploratory dynamism guarantees it a strong presence in each of the technologies conceivable for the future.

1.2.7. *The need for multinational coherence in energy policies: the case of difficult European convergence*

Concluding this section, it is worth considering the collective dynamics of countries beyond tropisms, not only on nuclear technology, but on the broader perspective of energy policy. The challenges posed by energy and the climate are planetary in essence and require international cooperation not only for the formulation of common goals, but above all, for the coherent and concrete implementation of associated actions. The Paris Agreement (2015) is an emblematic example of a convergent understanding of the climate issue; it constitutes a concrete step, as it invites each country to formulate its commitment in terms of measures that will contribute to the reduction of greenhouse gas emissions. But in virtue of its nature, global convergence which cannot be subject to a common authority will be a long and difficult road.

At the heart of a political space allowing for more sophisticated legislative constructions, Europe has already demonstrated the possibility of building an integrated and effective policy in the past. Today, it is confronted with technical and political complexities limiting the coherence and performance of its energy policy.

The European construction was born by the act of pooling energy resources, both for the dominant resources at the moment, namely carbon (1952) and for the more prospective resources, such as nuclear energy (1957).

Act 1 of European nuclear power was marked by the Euratom Treaty, which came into force in 1958. This political act directly translated a strategic perception of the vital importance of energy for economic and social development, the need to limit oil dependence (in relation to the Suez crisis in 1956), and to establish sovereignty in a balanced world. The treaty was drawn to consolidate a community of interests actively engaged in the "development of a powerful nuclear industry", which would "provide extensive energy resources" (preamble to the treaty). "It shall be the task of the Community to contribute to the raising of the standard of living in the Member States and to the development of relations with the other countries by creating the conditions necessary for the speedy establishment and growth of nuclear industries" (Article 1 of the Euratom Treaty; European Union 2010). The spirit of the negotiation of the treaty was based on two visions, one relating to the development of nuclear power as a means of controlling the future of European energy, and the other aiming at a major step toward a truly federal European integration. The Euratom treaty established community instruments for the ownership of materials, for the establishment of a conformity control system, for the shared programming of R&D and for a single market and coordination of industrial investments. Regarding these last two points, the Euratom Treaty was a step ahead for the EEC.

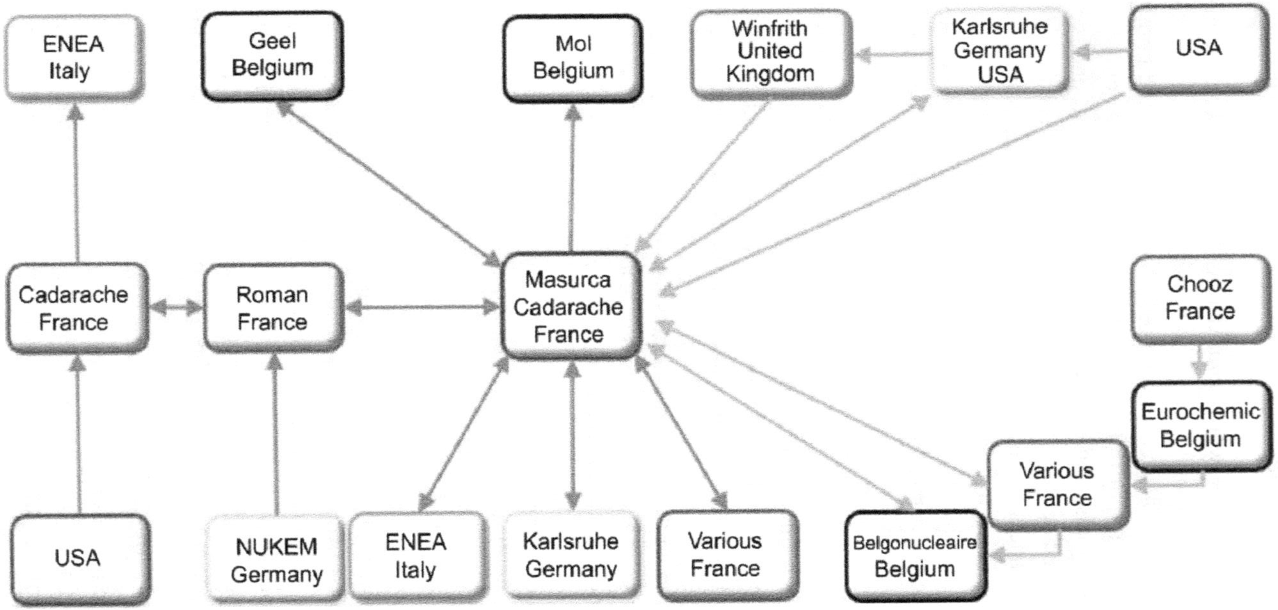

Figure 1.3. *Illustration of European collaboration in the development of nuclear technologies for the 1960–1970 decade. The diagram reflects the intensity of the uranium and plutonium flows exchanged between the countries (illustrated by different colors) within the framework of experiments carried at a facility, called MASURCA (Cadarache research center in France). For a color version of this figure, see www.iste.co.uk/percebois/nuclear1.zip*

Figure 1.4. *Illustration of the assembly of fuel rods in the tubes configuring the MASURCA core. For a color version of this figure, see www.iste.co.uk/percebois/nuclear1.zip*

This political framework and the will of the nations effectively enabled the rapid take-off of nuclear science and technology in Europe, encouraging the growth of a powerful industry as required by the treaty. Examining nuclear material flows (illustrated in the case of the MASURCA research facility in Figures 1.3 and 1.4 and Box 1.2) in the R&D field from the 1950s to the 1970s, brings out a vertiginous image illustrating continuous exchanges between countries, multiple recycling operations for the reuse of materials, with the goal of irrigating experimental exploration programs (at a simply unimaginable rate today).

Cooperation between States

In the 1960s and 1970s, cooperation between the United States and Europe was strengthened in the field of fast-neutron reactors (FNR). Highly enriched uranium and plutonium for civilian use were then rare, limited materials in research programs.

In 1964, the United States and Euratom reached an agreement for the supply of plutonium not only to France or Germany, but also to Italy, the Netherlands and Belgium. At the time, different circuits were used to supply the various research reactors: SNEAK in Germany, ZEBRA in the United Kingdom and MASURCA in France. Because of these three very low power research reactors, it was possible to explore and characterize the neutron behavior of fast neutron reactor cores. In addition, another part of the plutonium required for this research project was obtained by the reprocessing of spent fuel in Belgium, which was drawn from the Franco-Belgian Chooz nuclear energy plant. A batch of material of British origin was also supplemented with American material, marketed via Germany, to develop additional fuel for MASURCA in Belgium (1977).

Similar examples concern other French and European facilities involving material transfer operations for research programs, inter-facility material loans and more generally, multiple recycling and processing which engages the infrastructure from several countries. These material flows reflect the vitality of research in this pioneering era, as well as the shared motivation of many European countries to take part in a common scientific and technological adventure, encouraged and reinforced by the Euratom Treaty.

The MASURCA research reactor

The MASURCA core made it possible to study the neutron operation of fast-neutron reactors at very low power (5 kW), offering great experimental flexibility. Often described as mechanical, its core accepted all sorts of simulation elements (fissile or fertile fuel materials, inert or absorbent reflectors, etc.) in various geometries such as 24 cm cylindrical rods, for example. The fuel rods, like the other highly diversified simulation elements making up the MASURCA core, were assembled in long tubes (3.80 m) loaded from below the core.

Box 1.2. *Cooperation between States for the development of nuclear technologies during the 1960–1970 decade*

Shortly after, the large-scale developments became motivated by national logics to the detriment of the treaty's initial spirit (France also wanted to develop military uses, while Germany defended its national industry).

Act 2 of European nuclear power followed the general evolution in the field, with a phase involving knowledge and infrastructure acquisition in the 1960s and massive industrialization in the years 1970–1980, as a response to the oil shocks. After major nuclear accidents and the liberalization of the electricity markets, concerns have refocused on the safety of facilities and waste management. The first Nuclear Illustrative Programs (NIP) sent the Commission's investment recommendations to the Member States in Europe, which were unanimously adopted.

At the turn of the year 2000, the operational and collective character of the NIP was totally erased in favor of the recognition of the diversity of national choices framed by safety and waste directives. Solidarity around a common goal quickly disappeared and the balance at this stage was built around the right of interested countries to use nuclear energy against the commitment of a safety guarantee for all European citizens (Senate 2000). Striding away from an energy policy, the dominant goal then became the proper functioning of the energy market, by liberalizing the sector (in particular by separating production, transmission and distribution activities). Historic nuclear power, sovereign by virtue of its history and the responsibilities it entails, is under tension according to this new paradigm.

At present, the Treaty on the Functioning of the European Union (2012) establishes energy as a competence shared by the community and the Member States, each state retaining its right to "determine the conditions for exploiting its energy resources, its choice between different energy sources and the general structure of its energy supply".

In this context, the European nuclear Act 3 is characterized by rapid growth in the deployment of wind and photovoltaic technologies, accompanied by a significant reinforcement of interconnections which now raises the question of the behavior of the electrical system at continental scale. The goals for minimizing carbon emissions, reflected in the establishment in 2005 of the Emission Trading System (ETS) market, as well as minimizing electricity costs, could have created a favorable environment for European nuclear power. But this was not the case, illustrating the gap between technical and economic rationality and political arbitrations integrating other issues.

To varying degrees, many European countries considered that nuclear energy would not be a building block of their future energy policy. At the same time, these

European Union countries were heavily subsidizing large-scale programs for equipping photovoltaic and wind power generators, thus creating stress on the electricity market (negative prices, reduction of low-carbon controllable energies, etc.). As an example of this technological change, hydrogen is often perceived as a new technological wave which will enable the decarbonization of new economic sectors, or even the compensation for intermittency.

On a technical level, this transition to new energy technologies, in Europe as in many regions worldwide, is structuring the evolution of the electricity mix. Some actors consider that photovoltaic and wind power technologies are the tool for a fossil-free and nuclear-free energy future meeting the needs of advanced societies. However, uncertainties remain significant when one appraises the orders of magnitude and overall behavior of an electrical system, which has become complex and nonlinear. The technical and economic effects of intermittency are of little consequence as long as the installed power of photovoltaic and wind generators remains below the average annual power of electrical demand.

Today, this threshold has been reached in Europe. But the compensation for intermittency beyond this threshold is becoming a major challenge not only in technical terms (network stability) but also economically (additional cost of compensation means) and strategically (dependence on associated resources and technologies). These factors – which will be clarified from experience in the upcoming decades – cannot but reinforce the need for nuclear energy, the only controllable and carbon-free technology apart from hydraulic energy (whose capacity is limited). For the time being, nuclear energy appears as an element of security amidst the systemic uncertainties induced by intermittency as well as a decarbonization accelerator of a European electricity mix, which is still carbon-fueled to a 20% extent.

At a political level, the European community has evolved from a goal promoting nuclear science and technology toward a formal neutrality regarding the technological choices of Member States. However, network interconnection, the integration of the electricity market and the growing significance of the European Union's regulatory and financial instruments technically impose a much-needed collective convergence when it comes to assessing the advantages and limitations of the different available technologies to meet the Union's goals. In practice, the political divide around technologies is significant and tends to exclude nuclear technologies from European financial instruments, such as the taxonomy of sustainable activities or the transition funds intended to help countries opt out of carbon.

The formal respect for the sovereign technological choices of the Member States has today become an insufficient notion. At present Europe is confronted with the need to deal with energy issues respecting a global approach. This marks a positive development toward greater European integration in relation to frameworks for action which can no longer be essentially national. But this imposes a collective clarification, on the one hand, of the goals assigned to an increasingly integrated electricity system and, on the other hand, a shared technical investigation of the technological scenarios meeting these goals. So, if the overriding urgency is decarbonization, then nuclear power is part of the solution alongside wind power and solar energy, as the UK government has stressed (HM Government 2020a). If, on the contrary, as is the case today in Europe, the goal is formulated by the vaguer notion of sustainability, then the debate becomes more complex, involving the decarbonization and denuclearization issues, and can ultimately have a negative impact on European energy and climate performance.

The climate emergency is only growing because neither Europe nor the world are currently on the right decarbonization trajectory. To deal with the tension between climate goals and maintaining the economic and social standing of societies, the controversy between technologies – traditionally structured around an opposition between "renewable" vs. "nuclear", itself a political marker – will have to evolve toward a debate fueled by technical and economic realities on how to integrate technologies into a mix consistent with political goals.

1.3. The trajectory of the French nuclear ecosystem, between continuity and rupture

In the global electronuclear landscape, France is often described as an exception. One may wish to inquire further about this French exceptional character, its reality and its possible foundations. From the viewpoint of its electrical mix, the French case is undeniably unique. In 2019, nuclear power accounted for 70.6% of the electricity production[5] (despite a slight 3.5% decrease), making France the most heavily "nuclearized" country (given the proportion of nuclear energy in its total electricity production[6]). In 2021, this production relied on 56 commercial reactors, spread over 18 nuclear energy plants. But, beyond these figures, the singularity of the French case is also related to its history and to the network of stakeholders on which the nuclear ecosystem is based.

5 According to the reports compiled by RTE in 2020 (https://bilan-electrique-2019.rte-france.com/production-totale/#1).

6 Note that in 2016 the United States were the country with the highest number of nuclear reactors in operation (99) for a total production of 804.9 TWh.

This section aims to provide a qualitative and dynamic description of the French nuclear ecosystem's trajectory. The purpose is to understand the developments of this ecosystem over time as well as its challenges for the future, apprehending its "technopolitical" background (Hecht 2004). How was the French nuclear strategy devised and implemented? What determined the choice of nuclear reactor technologies to be deployed? How did these technologies concatenate in time? Which players is the French nuclear ecosystem based on? How have their roles and relationships evolved over time? Under the influence of what national or world events? Finally, how to describe the current situation of civil nuclear energy in France and its possible future? These are all questions that will be addressed in this section. As will be shown, the history of civil nuclear energy is neither linear (Torres 2016) nor a natural succession of ever-improving technologies. On the contrary, it is marked by conflicts, struggles for influence, reconfigurations and an alternation between phases of intense development and sudden slowdown.

Today, nuclear power faces major challenges, particularly in France, and the questions about its future have never been so poignant. In the early 2000s, however, after a decade of low activity in the design and construction of new reactors, a revival seemed to prevail, raising a surge of optimism associated with what some authors have come to describe as "nuclear renaissance" (Nuttall 2004). In France, such a movement seemed to be embodied by the active involvement in the Generation IV International Forum (GIF) launched by the American Department of Energy (DOE) in 2000 to re-encourage R&D efforts in the field of civil nuclear energy, leading to the inauguration of the ASTRID[7] design project in 2010, allegedly the first French "technology demonstrator" of a so-called "Generation IV" reactor. In 2007, this revival was materialized by the initiation of the construction site for the first French EPR[8] reactor in Flamanville. In recent years, this nuclear renaissance seems far more unpredictable. The Fukushima accident from 2011 reactivated questions about civil nuclear energy and its safety on an international scale. In France, contrary to what was expected, the flagship projects of the 2000s did not follow the trajectories announced: the EPR, whose commissioning was initially announced for June 2012, was still not operational by 2021, and the ASTRID project officially came to a halt in August 2019. At a political level, in October 2014, the National Assembly voted Article 1 of the law relating to the "energy transition for green growth" providing for the reduction of the share of nuclear-generated electricity from 75% to 50% by 2025, fulfilling one of the

7 Advanced Sodium Technological Reactor for Industrial Demonstration.

8 European Pressurized Reactor. A first EPR reactor had already been under construction in Finland since 2006.

promises made by François Hollande during his electoral campaign. Very recently, two reactors were definitively shut down in Fessenheim.

As these events illustrate, the trajectory of the nuclear ecosystem seems discontinuous, marked by phases of revival or decline. By retracing this trajectory, this section seeks to offer key elements to understanding the particular issues and challenges faced by the French civil nuclear industry. Exogenous factors, such as nuclear accidents or global warming and the efforts to be made in relation to the energy transition, as well as endogenous factors regarding the French sociotechnical ecosystem and its changes over time, will be discussed in the following paragraphs.

1.3.1. *The origins of French civil nuclear energy*

1.3.1.1. *The emergence of two key players: the French Atomic Energy Commission and EDF*

The history of French nuclear energy is said to have begun with the construction and commissioning of the ZOE atomic pile[9] in 1948, the first experimental nuclear reactor in France, which was met with great enthusiasm in post-war France. The official statements and newspapers of the time attest to this, describing ZOE's success as "a decisive milestone in France's civilizing mission" (Hecht 2004, p. 6), praised for its glory (*op. cit.*). The emphasis was then placed on the medical and research applications expected from such an innovation, more than on its third goal, the military application. However, while it was true that ZOE granted the production of radioelements for hospitals and laboratories, its exploitation offered scientists the opportunity to isolate the first milligram of French plutonium. This experimental reactor played a crucial role in validating the technical processes for producing plutonium, which could potentially be used for military purposes. Thus, from the very beginning of nuclear power, the boundaries between civil and military nuclear energy were thin, to say the least.

Anxious to "make France great again", General de Gaulle supported the creation of the French Atomic Energy Commission (*Commissariat à l'énergie atomique*, CEA), decreed in October 1945. The scientist Frédéric Joliot-Curie obtained the post as High Commissioner, whereas the former minister Raoul Dautry was appointed General Administrator. Since its creation, the CEA was entrusted with the role of conducting research on atomic energy and its scientific, industrial and defense

9 The Z stands for "zero", in reference to the reactor's very low power. The O stands for "uranium oxide", the fuel used in the pile. Finally, the letter E refers to the moderator made up of "heavy water", whose role is to slow neutrons down. This has the effect of increasing the probability of fission in the reactor core.

uses[10]. It also obtained a special status specifying that it exclusively reported to the Prime Minister. Since the very beginning the Parliament was excluded. The CEA stimulated a process of reflection and development in the nuclear sector[11], inviting numerous debates among scientists, engineers and political actors as to their advantages and disadvantages, depending on their purpose (to produce electricity for civil purposes and produce plutonium for military purposes). Given the availability of resources in France (mainland and the colonies), uranium and graphite were then preferred as fuels and moderators for the first reactors. This choice was included in the 5-year plan (1952–1957) for the development of atomic energy.

In the 1950s, alongside the CEA, another player gradually became essential in the nuclear adventure, *Électricité de France* (EDF), at the moment the nation's main electricity supplier[12].

In the initial days of the history of French nuclear energy, the CEA (mostly a research body) brought its technical know-how to EDF, who would be trained in mastering nuclear energy. In return, EDF would put its industrial competencies at the service of the CEA. This signaled the beginning of a fruitful a priori 15-year collaboration, marked by the deployment of the earliest "Generation I" French nuclear reactors. Decisions pertaining to the nuclear program were formally made by the executive power, following the advice of the PEON[13] committee, created in 1955. At first it had a lesser role, the decisions being chiefly negotiated between the CEA and EDF or within each organization. Between 1956 and 1966, construction work began on a new gas-graphite reactor system (*Uranium Naturel Graphite Gaz*[14] [UNGG]) Added to these was the commissioning of a Heavy Water Gas Cooled Reactor (HWGCR) at Brennilis and an American technology Light Water Reactor (LWR) at Chooz.

10 Available at: https://www.cea.fr/Pages/le-cea/histoire-creation-CEA.aspx.

11 This typically French term (*filière*) designates the entire nuclear industrial chain, upstream and downstream of the cycle, including reactor exploitation. For simplicity, the term *filière* can also describe a reactor technology used to produce nuclear energy (which differs depending on the type of fuel and moderator).

12 In 1946, a nationalization law was passed bringing together private companies within a single public electricity company, EDF, responsible for its expenditure to the Ministry of Finance and for its development program to the Ministry of Industry (Hecht 2004, p. 51).

13 PEON (*Commission pour la production d'électricité d'origine nucléaire*) was an advisory committee bringing together senior managers from EDF and the CEA, ministerial and senior civil servants and industrialists related to the nuclear sector (Hecht 2004, p. 262).

14 These were the G1, G2, G3 reactors at Marcoule (commissioned between 1955 and 1960), EDF 1, 2 and 3 (1963–1966) at Chinon, EDF 4 and 5 (1969, 1971) at Saint-Laurent and Bugey 1 (1972). All these reactors are now shut down and decommissioned.

1.3.1.2. *The war of the systems episode*

However, as EDF improved its skills in the management of its nuclear fleet, the first dissonances were heard as to the competitive nature of gas-graphite reactors. In 1967, the Horowitz-Cabanius report, named after the respective leaders of the CEA and EDF, brought to light the tensions between the two organizations, tensions that would last over 3 years and would be analyzed by G. Hecht as "the war of the systems". This was rooted in the State's desire to increase its nuclear electricity production capacity, not only to meet a growing demand, but also to strengthen energy independence. The question came down to choosing the technology (or technologies) to be retained for the future development of civil nuclear energy, in a context of growing uncertainty, around which the stakeholders tensions were increasing.

From 1967 to 1969, each organization sought to promote its preferred technology: while the CEA defended the gas-graphite reactor technology, which it had designed itself, EDF advocated the use of pressurized water reactors (PWR), American technology, developed by Westinghouse. During the conflict, each opponent argued its position. In the voice of Cabanius, EDF defended an economic-industrial logic consistent with its own definition of a public service company: it came down to the question of supplying electricity to the greatest number and at the lowest cost, including French companies. EDF played the dual role of customer and supplier to French industry, claiming to strengthen the French industrial fabric and its international competitiveness (Hecht 2004).

EDF insisted on the inability of gas-graphite reactors to provide sufficient power, a major issue in the context of the energy independence policy pursued by France in a parallel manner. In the voice of Horowitz, the CEA argued for the success of gas-graphite technology, the experience acquired and the importance of a national technology for promoting France's glory. According to G. Hecht, this "war" was fueled by the technopolitical regimes underlying each organization: "nationalized" for EDF, "nationalist" for the CEA. A major point in the definition of these regimes is precisely the way in which each opponent characterized the relationship between the "technical" and the "political".

In line with previous decades, the CEA strove to maintain close links between technology and politics, with the goal, for Horowitz, of "addressing political issues directly" (Hecht 2004, p. 254).

On the contrary, at the time EDF declared "politics [...] [was] none of its business" (*op.cit.*), expressing a desire to separate technique and politics in favor of a more overtly economic rationality.

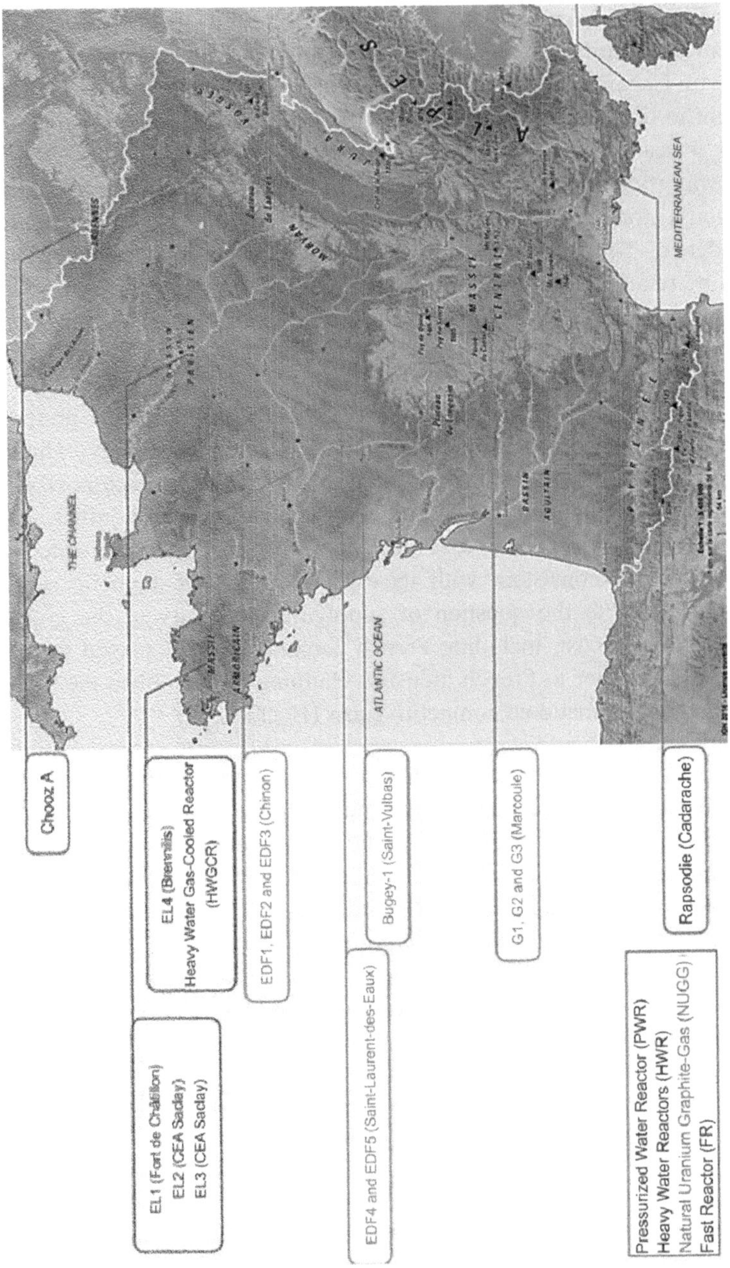

Figure 1.5. *The development of first-generation reactors. For a color version of this figure, see www.iste.co.uk/percebois/nuclear1.zip*

During this period, the role of the PEON commission "gained in subtlety and complexity" and "carefully nurtured the status as objective arbiter" in the debates (Hecht 2004, p. 263). After a 3-year "war", an agreement was reached between Marcel Boiteux from EDF and Robert Hirsch from the CEA. In 1969, the executive power – on the dictates of the new president, Georges Pompidou – decided in favor of PWR technology, on the basis of primarily economic and industrial arguments. With this decision, the French nuclear industry shifted from the "pre-industrial stage, essentially involving experimentation, research, trial and error, to an industrial stage, where the concern for rationalization and optimization became permanent" (Garcias 2014, p.79). A few years later, the CP0 program started, marking the beginning of a massive deployment of civil nuclear energy in France.

1.3.2. *French nuclear power in the industrial age*

Despite its long history, the French nuclear program accelerated at the beginning of the 1970s, with the nationalization of Algerian hydrocarbons (1971), and the first oil shock (1973). It gave the French government the opportunity to reaffirm the relevance of nuclear power for national independence in relation to oil-producing countries. In 1974, the Messmer plan (named after the Prime Minister at the time) called for the launch of a vast program to build nuclear reactors. From 1974 until the end of the 1990s, a succession of programs resulted in the construction of a nuclear fleet comprising 58 PWR-type reactors at 19 geographical sites. During the 1970s and 1980s, construction sites around the world followed one another at an unprecedented rate (*op. cit.*). Not only did the massive deployment of nuclear energy in France[15] signify a technically daunting task; the challenge was also socioorganizational: it urged training the staff to operate those power plants. In short, it required producing highly specific skills, on a large scale, in a brief period of time. But it was also necessary to propose an adequate organizational model to support such a program: enriched by its own experience in thermal and hydraulic power, EDF proposed a "stepwise" construction model, in other words, done "in a series of identical units" (Taccoen 2003, p. 102). The company relied on a strategy inspired in design replication and identical construction principles, with a view to optimization and standardization.

The French institutional and industrial context, seeking to transcend a merely prototyping logic, promoted an "integrated" model between operator and project coordinator for the design/construction of the nuclear fleet. This model was

15 It should be noted that this deployment was not devoid of clashes and disputes, but on the contrary, sometimes inspired violent opposition. On this subject, see Marcel Boiteux's testimony in the Arte documentary "Nuclear: nothing to report" ("*RAS nucléaire – Rien à signaler*", 2009).

embodied in EDF's dual role both as a customer (operator) and as an equipment supplier (turnkey contractor). For Marcel Boiteux, this dual status of the company would be "another success factor, specific to the French case; it is the customer himself/herself – EDF in this case – who acts as its own industrial architect and provides general engineering, instead of an independent architect as in the United States, or the nuclear boiler supplier itself as in Germany. Thus, EDF knows the "why, the how..." (Boiteux 2008, p. 40). To a large extent, this situation made it possible to curb the inclinations of the engineering teams who wished to continuously refine the design of power plants and contain innovation to the benefit of learning rates and economies of scale: "the supplier's (or project coordinator's) new and brilliant ideas being neatly tucked away in folders to be exploited later" (*op. cit.*). This deployment was accompanied by major changes in the nuclear ecosystem, modifying the respective positions and roles of EDF and the CEA, but also seeing the rise in power of stakeholders who had until then remained in the background, starting with Framatome[16].This Franco-American company had developed a certain expertise in the design/construction of PWRs in the United States. The French program was therefore partly based on the expertise of American engineers.

Finally, after about two decades, the Messmer plan had reached its main goals. This development was deemed "successful" by the author for at least three reasons: its scale (from almost 0% to 80% of electricity produced), its speed (program deployed over less than 20 years) and its systemic nature (the deployment of a global industrial capacity, covering the entire activity chain, from fuel extraction to waste treatment, including the production and distribution of electricity, the construction and decommissioning of production units). Several reasons can account for this industrial success: "the adoption of a technology already explored in the United States, partly to the detriment of national independence, but to the benefit of significant savings in R&D and learning; the choice of a technology-stepwise replication strategy; a centralized industrial organization around a key player (EDF), conditioning replication and minimizing coordination costs" (Garcias 2014, p. 82).

1.3.3. *A story within a story: sodium-cooled fast reactors (SFRs)*

Following the decision to massively equip France with PWRs, a solution advocated by EDF, the place and role of the CEA in the French nuclear ecosystem began to change. The commission lost the technical and industrial leadership it had held in the construction of nuclear reactors during the 1950s and 1960s, to the hands of EDF. At the end of the 1960s, it had to "give up the leadership in the construction

16 Franco-American Atomic Society.

of power plants" (Torres 2016, p. 83), while ensuring it maintained the scientific and technical expertise in that field. It strategically repositioned itself by becoming the world leader in the overall management of the nuclear cycle: upstream (with the extraction, conversion and enrichment of uranium) and downstream (by developing key skills in the processing and recycling of spent fuel). In particular, those activities were handled by Cogema[17], a 100% subsidiary of the CEA created in 1976.

The construction and operation of La Hague plants (UP2 then UP3) played a central role in the implementation of the French strategy for the recovery of nuclear materials. In parallel with the deployment and operation of pressurized water reactors (PWR, known as "Generation II" reactors), the CEA worked on a possible symbiosis between PWRs and so-called fast-neutron reactors (FNRs)[18], offering the possibility of eventually reaching a closed nuclear fuel cycle. The concept of sodium-cooled fast reactor (SFR) was then privileged. The SFR system embodied the consensus between the supporters of PWRs and gas-graphite reactors, and repositioned the CEA in the field of reactor design/construction. Between 1967 and 1984, three reactors based on this technology were commissioned: the Rhapsodies, a very low power reactor, the Phénix, a small-scale experimental reactor of 250 MW and the Superphénix, an industrial demonstrator of 1,200 MW[19]. Through the Phénix, and then the Superphénix, the CEA committed itself to developing its expertise in the SFR sector. This scientific, technological, human and financial effort was part of an elaborate 1970s strategy planning for and organizing the replacement of the PWR fleet still operative today by an FNR fleet capable of handling the thorny issue of spent fuel from PWRs, as well as (and mostly) enhancing the recovery of uranium 238[20] (Tillement et al. 2014). Since the launch of the electronuclear program, the CEA's strategy has incorporated the question on how to manage nuclear waste and materials. According to its promoters, the interest and legitimacy of the FNR system are largely based on its ability to recycle the plutonium produced by the current PWR fleet. This contributes to legitimizing the

17 General Nuclear Materials Company.

18 Unlike PWRs, a fast-neutron reactor (FNR) uses neutrons emitted by fission which are slightly slowed down. They are described as fast because their kinetic energy is high. In the context of the French fleet, the only technology having worked relied on the use of liquid sodium as a heat transfer fluid.

19 None of these reactors are currently in operation. The Superphénix was shut down in 1997 and the Phénix was operational until 2010.

20 Unlike PWRs, FNRs have the ability to efficiently upgrade uranium 238 by transforming it into plutonium and fissioning it. This ability has led the promoters of this system to dream of an (almost) indefinitely available source of energy, making it possible to break free from mining extraction for good.

fact that in France plutonium is not considered as "waste" but as a recoverable matter, having been reprocessed for many years in La Hague plants.

The shutdown of the Superphénix, by a Government Resolution endorsed by Prime Minister Lionel Jospin in 1997 – against the will of the CEA and EDF who considered it unjustified both from a scientific and technological viewpoint – represented a "hard blow" for the industry. Not only did this hinder the development of SFRs and the possibility of attaining a closed cycle, but it also jeopardized the expertise gained in the field. Following this decision, the sodium system was abandoned for some time, but the reprocessing of spent fuel, so as to remove plutonium, continued and constituted one of its major successes. This disengagement only lasted for a short time. The "Generation IV" forum launched in 2000 by the American DOE was a great opportunity for the CEA to reconnect with SFRs, as did the law relating to the management of radioactive waste (Bataille's law of 1991, then of 2006)[21].

In the French context, the strength of the direct link between decision-making processes relating to "GEN IV" design and the management of nuclear waste stems from the initial industrial strategy. Placed in this historical context, it is clear that the "GEN IV" project was an opportunity for the CEA to relaunch its skills in the field and to prove the quality of its strategic vision for the organization of the French nuclear industry as a whole. This renewed interest was materialized by France's commitment to the study and construction of a SFR prototype[22], named ASTRID and piloted by the CEA. The determination to place the ASTRID demonstrator project in line with the Phénix and the Superphénix reactors and the magnitude of the investment made (around 1 billion euros for the design studies) show to what extent it was crucial for the nuclear players (namely, the CEA) to once again become a leader in SFR technology and more broadly to fully master the closed nuclear cycle. Later in the text the particular case of the Generation IV, and the ASTRID reactor project will be discussed.

1.3.4. *The "white" years: the gradual decline of new projects*

The 1980s were marked by a slow progression in the nuclear fleet, as the design of power stations evolved. Inspired in the basic model developed by Westinghouse, EDF and Framatome engineers would gradually make improvements, increasing the

21 This law provided for three main radioactive waste management strategies: surface or sub-surface storage (a few meters to a few tens of meters underground), deep geological burial (with the CIGEO project) and waste reduction via incineration in fast reactors.

22 The beginning of the construction for this demonstrator was originally planned for 2020.

power and safety of reactors by integrating "experience feedback" from those previously installed. This period also saw EDF and Framatome gradually free themselves from the supervision of Westinghouse. The P'4 tier was presented as a French version of Westinghouse's P4 design. As for the N4 tier, it was the first to be presented as a totally French concept, and the four currently operating reactors of this type (at Chooz and Civaux) do not depend on the American license. But the decade of 1980 also witnessed a gradual decline in reactor construction, as the saturation line of the fleet's production capacities drew nearer.

Even if strongly associated with national contingencies (the risk of overcapacity), this plateau effect was part of a more global movement characterized by a sharp slowdown in the deployment of civil nuclear energy in the world. This phenomenon had a dual origin: it was explained both by the fact that some of the most active nuclear energies (France, Japan) had reached a critical production capacity before the renewal horizon of such capacities, and by the opting-out of other powers from nuclear technology, mainly due to growing safety concerns. The first of these great nations was the United States, which suspended its program following the accident at the Three Mile Island (TMI) power station on March 28, 1979. While most of the projects in progress had by then been completed, no new projects were undertaken thereafter. But the real turning point came with the Chernobyl accident (1986), far more serious than TMI in terms of its human, health, economic and even environmental consequences. Its impact on reactor construction around the world was particularly dramatic. In fact, the number of operating reactors had reached its expansion peak that same year. Since then, and until the present, the "installed power", that is to say, the total electricity production capacity of the world's reactor fleet, has continued to stagnate, with the few reactors entering into service each year (an average of 4 or 5 per year), only compensating for the shutdowns of old reactors. This global stagnation has been aggravated by the total withdrawal of countries like Germany.

After a period of "euphoria" during the 1970s and the beginning of the 1980s, the 1990s corresponded to a period of low activity, sometimes qualified as "nuclear winter"[23].Unlike other countries, France did not completely turn its back on nuclear energy, though its production slowed down: during the 1990s, the four N4 type reactors made it possible to maintain activity, but the players in the sector feared soon finding themselves with no other prospect than to operate the 58 existing reactors and dismantling the reactors which had been shut down. In addition to this

---

23 This expression was notably used in a parliamentary report on the safety and security of nuclear facilities, written in 1998: "Control on the safety and security of nuclear facilities", Claude Birraux (Deputy), Jean-Yves Le Dréaut (President) and Henri Revol (Vice-President).

drastic decline in engineering and construction activities, there were other disappointments, in particular, the shutdown of the Superphénix.

By the end of the 20th century, the Messmer plan had successfully been completed: France had 58 reactors in good working condition, capable of supplying a large part of the electricity the country needed for a few more decades. But the whole industry was then plagued by doubts regarding its future.

1.3.5. *The painful emergence of a "nuclear revival" in France: projects, organizations and competencies*

1.3.5.1. *2000–2010: the idea of a nuclear renaissance in France*

At the beginning of the 2000s, the sky seemed to be clearing up for the nuclear industry. The whole industry started to believe in a bright future again, after years of expectation and dashed hopes. The sector's belief in a possible "renaissance" was based on the conjunction of several elements:

– The aging of "historic" plants in countries which had not explicitly declared their desire to give up nuclear energy (mainly the United States, France and Japan) could call for "replacement" constructions.

– Awareness of the growth potential represented by so-called "emerging" countries, such as China, India or Brazil, some of which had already embarked on a civil nuclear program and could pursue a massive equipment strategy, summoning the expertise from the large Western companies.

– A significant displacement of the objects of ecological concern, with the rise of the "global warming" issue. Associated with the depletion perspective of world hydrocarbon reserves, it tended to discredit the use of fossil fuels as a source of electricity production.

In this context, nuclear energy appeared as a source of energy which – as long as it was controlled in terms of safety – represented a "clean" and "sustainable" alternative to the combustion of fossil fuels (highly polluting and too dependent on commodity price fluctuations). In 2004, Corbel summarized it as follows: "The leaders of the nuclear industry have good reason to hope for a recovery. The current energy system, based on fossil fuels, is not satisfactory [...]. Despite their growth, no renewable energy today has the potential to replace fossil fuels. In this view, the absence of direct greenhouse gas emissions, and the relative diversity of uranium resources are incomparable advantages of nuclear technology. The potential for recovery is therefore significant" (Corbel 2004, p. 157).

This "hope" in the potential for recovery crystallized during the first years of the 21st century in the rhetoric of the "nuclear revival"[24]. At that time, the experts quantified the number of new reactors likely to be commissioned in the coming decades, integrating the prospects for growth in energy demand (particularly related to emerging countries) and the nuclear industry's share for satisfying such demand. They estimated that on average, the growth in installed nuclear energy would be around 75% in 20 years, increasing from 390 GW to 690 GW by 2030. This hope for an industry revival explains why, despite the low number of projects underway since the mid-1990s, very few players actually opted out from the market, preventing a real decline in the nuclear sector (Corbel 2004). Moreover, the period was marked by a major restructuring and regrouping of nuclear stakeholders.

In France, the early years of the 21st century were characterized by the greatest transformations. In 2000, Framatome and the nuclear subsidiary of Siemens merged into an entity called Framatome ANP. In 2001, Framatome and Cogema, already related via the CEA-Industry structure, were merged under the banner of the new French nuclear conglomerate: Areva. In 2006, Cogema became Areva NC, an entity in charge of the fuel cycle, whereas Framatome ANP became Areva NP (Nuclear Power). For its part, EDF underwent a major institutional change between 2004 (transition to the status of public limited company) and 2005 (initial public offering, the State retaining a large majority of the capital, up to 84%). This change in status enabled EDF to change its scope of action, and to now consider the possibility of being a fully fledged nuclear player at a global scale, beyond the mere role as technological support for an operator. The former public company with the State monopoly was called upon to become actively involved in the "nuclear revival".

By the end of the 2000s, the French industry seemed prepared to give life to the wish for a "a nuclear revival". ASTRID and the Flamanville EPR had to attest to this evolution. The goals assigned to them were particularly ambitious. The construction onset for ASTRID (whose design was monitored by the CEA) was initially scheduled between 2020 and 2025. The construction of Flamanville EPR, under the responsibility of EDF after the basic design by Framatome and Siemens, and then by Framatome alone, was expected to last 5 years (between 2007 and 2012).

The challenge for ASTRID as for the Flamanville EPR was not to take over from a well-established sector, but to give concrete expression to the revival wish of an entire industry. A few years later, the target had only been partially achieved. By 2020, Flamanville had still not been commissioned, and the construction of a possible demonstrator for a Generation IV reactor had been postponed until the second half of the 21st century.

24 *Nouveau nucléaire*, in French.

1.3.5.2. *"Showcase" projects struggling to keep their promises*

Within a few months, between the summer and autumn 2019, the difficulties experienced by the French nuclear industry came to light, with two announcements. The first one, in the summer, concerned the ASTRID reactor project, which was brought to a halt after 10 years of studies under the aegis of the CEA. The second one, in the fall, involved the so-called "Generation III" EPR at Flamanville: the presentation of a report commissioned by the State to industrialist Jean-Martin Folz was interpreted as a severe criticism on how the project had been conducted, by then 7 years behind schedule. Questions multiplied in the press as well as in the political sphere: could the French nuclear industry, renowned for its international excellence and the incomparable experience of its companies, still "rise to the occasion"? Was it able to meet the challenge of giving France, via its ability to build and operate reactors, pride of place in the decarbonization of electricity production, a major challenge for the 21st century? Or in reducing dependence on uranium? Could it perpetuate the "French exception" and its highly nuclearized electricity supply via the renewal of the existing stock?

These two events sounded like a major warning, tarnishing the image of the main organizations concerned, among which EDF, ex-Areva (now split between Orano and Framatome) and the CEA were seriously called into question. The question was no longer whether the continuation of the French nuclear adventure was desirable (a debate which had already occupied the forefront of political and media controversies for thirty years), but now examined whether such pursuit was possible at all, at what price, and under what conditions.

Coming to grips with the difficulties encountered during the ASTRID and EPR projects, despite their differences, is key to understanding the situation in which the French nuclear industry found itself at the beginning of the 2020 decade. But this understanding requires a drastic step back from a superficial and short-sighted reading of events. It makes it necessary to update the highly singular status of these two projects, the conditions under which they were initiated and implemented, and their place in the historical trajectory of French nuclear energy.

The decision to bring the ASTRID project to a halt was made in 2009 as part of the "investments for the future" program (also known as the "big loan") under the Sarkozy presidency and seemed to obey a simple change of priorities by the head of the State. The request for a halt formulated by the Ministry of Ecological and Inclusive Transition was justified in these terms: "the need for a demonstrator and the deployment of fast-neutron reactors were not useful, at least until the second half of the 20th century". The nuclear sector was thus called upon to concentrate its efforts on the most urgent achievements: "the industrial sector must also prioritize the success of the Generation III EPRs [...]. In this context, the construction onset for an industrial-scale demonstrator is neither necessary nor appropriate". Once considered highly strategic issues (back in 2009), the technological innovations required to "close the cycle" – or at least to lower uranium consumption and increase fuel recycling – now seemed as if they could be put on hold for several decades. What can explain this change in perspective?

The most obvious reason was put forward only a few weeks later on the occasion of the report submitted by Jean-Martin Folz to Bruno Le Maire. The latter confirmed to the public authorities, and more broadly to the public opinion, that the series of delays and problems encountered during the EPR project in Flamanville could not be interpreted as simple accidents along the way: they were indicators of much deeper underlying issues. Browsing the plan of the report provided an immediate overview of the content of the criticisms made: "a host of negative events", "an unrealistic initial estimate", "inappropriate project governance", "struggling project teams", "insufficiently advanced studies at launch", "unsatisfactory relations with companies" and, finally, "a general lack of competencies".

How could problems of this order have affected an entire industry once held up as an example not only for its ability to implement the Messmer plan, but also more generally, praised for the absence major nuclear accidents, or for controlling electricity costs in France? The Minister of the Economy, Bruno Le Maire, reacted to the submission of the Folz report as follows: "these delays are not acceptable and these repeated difficulties negatively impact the nuclear sector, which must pull itself together and do so quickly"[25]. This reaction has to be weighed against the trajectory of the French electronic nuclear industry: a feeling of failure is only ever relative to standards of judgment resulting from the knowledge of past achievements, at the same time that it is proportional to the magnitude of the hopes and expectations placed in a project.

1.3.5.3. *Growing uncertainty about the deployment of a "nuclear revival"*

During the 2000s and 2010s, the French nuclear ecosystem was once again the scene of tensions, wavering between the main players in the sector. In the advent of the "nuclear revival", this put to the test the industrial organization capable of effectively taking charge of the hoped-for "second wave" of nuclear energy reactor construction as well as its main stakeholders.

These tensions were certainly far from new: the history of the sector had already been marked by the "war of the systems" episode in the 1960s, seeing the triumph

25 Interview at RTL on April 28, 2019.

of EDF with its rationalizing conception of the nuclear industry. But, with hindsight, these tensions then seemed to take place within the framework of a relatively classic interplay of actors, within a clear distribution of roles, under the aegis of a State simultaneously acting as arbiter and strategist. The central place occupied by the State in French nuclear affairs has been highlighted as a decisive factor in the success of the nuclear energy program: for Grubler (2010), nuclear power has been more successful in France than elsewhere (especially in decentralized nations like Germany or the United States) precisely because of the political centralization of France and its interventionist tradition, adapted to an industry in need of stability, continuity and predictability a priori. In Finon and Staropoli (2001), "strong political support, a State electricity monopoly with substantial engineering resources, a highly concentrated electromechanical industry, an influential R&D agency, operating within the framework great regulatory stability, and effective coordination, resulting from long-term organizational arrangements" were all ingredients behind the success of the nuclear energy program.

Without having totally disappeared, some of these ingredients seem to be lacking today. As Yves Bréchet pointed out in a column deploring the abandonment of ASTRID, the State seems to have largely given up its role as a strategist (Bréchet, 2019)[26].This desertion, albeit partial, of the arbitral and strategic role of the State, greatly destabilized organizations. The sector seemed to have lost its backbone, giving free rein to the multiplication of situations in which the role and status of the actors was uncertain, or lacking legitimacy. On the occasion of the EPR project from Olkiluoto in Finland[27], Areva wished to experiment the supply of a turnkey plant to the Finnish operator, seeking to position itself no longer as a partner but as a competitor of EDF on its historical missions. Engaged in a new rivalry, Areva and EDF competed with sensational announcements as to their respective EPR projects. In 2020, a report from the Court of Auditors on the EPR sector argued that "the rivalries between the two national public groups, not arbitrated by the political authorities of the time, resulted in a dangerous escalation for the French nuclear sector". The challenge of this bidding was to reach international markets through the promise of powerful, reliable reactors, built quickly and at a lower cost. The construction deadlines announced (4.5 years and 5 years) quickly proved to be unrealistic and backfired for the builders, who subsequently suffered numerous setbacks in international calls for tenders. Long setbacks followed for Areva, meanwhile split between Orano and Framatome, a boiler design subsidiary, finally taken over by EDF who tried to save what was left.

26 Available at: https://revue-progressistes.org/2019/09/22/larret-du-programme-astrid-une-etude-de-cas-de-dis-parition-de-letat-stratege/.

27 This project, which preceded that of Flamanville by 1 year, is now complete.

In the same vein, although on a much lower scale of tensions, the ASTRID project also experienced its share of uncertainties regarding the respective roles of the organizations involved in the design. While the CEA took on the role of project coordinator (thanks to its experience with the fast-neutron system), criticisms were made as to its legitimacy to occupy such a role in a project originally intended as a "technological demonstrator", with a view to industrialization. In fact, the CEA was a research organization rather than an industrial company.

Although EDF was announced as a key player in the official documents in the project consortium, its real interest for seeing the project through remained largely ambiguous during the ten years of the program. It must be said that those same 10 years correspond to the escalation of the ambushes encountered during the realization of the Generation III Flamanville EPR project, which monopolized most of the electricity supplier's attention. If EDF was reluctant to become fully involved in an industrial program on a Generation IV reactor, it was certainly because the public company was primarily concerned with defining a clear medium-term strategy for the renewal of the existing fleet. However, ASTRID contributed little to this, as this project aimed to prepare the "next move". Despite its excellent motivations (closing the cycle, developing a new solution at a global scale, etc.), ASTRID struggled for survival in the face of the urgency for renewing the fleet, a scenario which still looked too uncertain. The only stable working hypothesis for EDF was that of a decrease to 50% of the share of nuclear energy in the French energy mix. On the other hand, the conditions under which 2035 targets can be achieved are still uncertain due to the aging of a number of reactors built during the 1980s. In 2021, the official position of the French government was to postpone the decision whether or not to build new EPRs (in a certainly "optimized" format), until the commissioning of the Flamanville 3 reactor. This means that no decision will be made until at least 2023.

This lengthening of the time interval between the launch of construction sites poses major management problems for EDF and, more broadly, for the sector as a whole. As stressed by the Folz report, and by other academic works (Garcias 2014), too great a gap between construction onsets hinders the capitalization dynamic of the know-how (both in terms of design and realization) to the point of eroding individual and collective competencies. This can be explained by the loss of habits, retirements which are not systematically compensated for, or the turnover within teams, a source of worry both for project managers and subcontractors. While the construction of the existing fleet had been built within a particularly tight period of time (the year 1981 alone saw the commissioning of 8 units in France), it positively impacted on economies of scale and prompted powerful learning effects. On the contrary, the second wave of reactor construction began in a radically different way.

At the start of the Flamanville EPR construction site in 2007, 16 years had elapsed since the companies in the sector had been summoned to start a new construction site (i.e. since 1991). As a result, some of the engineers, technicians and workers or subcontractors involved in the project were lacking the significant experience in the field of nuclear construction.

This "widespread loss of competencies" pointed out by the Folz report was not compensated for, despite massive hiring by EDF during the first years of the EPR project. The arrival of a large number of recently graduated engineers to EDF's design departments led to tensions due to the need to train young professionals "on the spot", a highly time and resource-consuming activity. Those inputs were not always available in such an emergency context, which was rather constrained by ambitious construction deadlines (Garcias 2014). This "learning wall" (*op. cit.*), a consequence of the intermittent nature of the work schedule of nuclear companies, was all the more complex to negotiate, as the organizations in question needed not only to rebuild lost competencies, but also to build new ones. It is virtually impossible to recreate the outstanding quality of the engineering departments in charge of the construction of the PWR fleet. The regulatory context has significantly evolved between the 1990s and our days, due to the strengthening of safety requirements and the birth of an independent safety authority. From the point of view of design methods, the landscape is also upset by the arrival of digitization, or the generalization of new approaches such as system engineering.

The challenges seem to be accumulating for the actors of the "nuclear revival". However, they must solve them if they wish to once again make nuclear energy attractive to the new generations of engineers, technicians or qualified operators. This attractiveness is an essential condition for the long-term maintenance and development of industrial capacities. This can only be sustainably built with stable prospects. Nuclear power was already a socially contested industry; now, with the difficulties encountered on "showcase" projects, it is the image of a declining industry. However, this decline is not inevitable: nuclear energy reactors can always be presented as an instrument in the fight against global warming, and the maintenance of a French nuclear industry as a protective measure against galloping deindustrialization. The fields of innovation, the technical and socio-organizational challenges to be met are numerous in the field (maintenance of the current fleet, better use of fuel, recycling and storage of waste, construction of smaller and more modular reactors, etc.). Despite the many difficulties related to the context in which it is initiated, the "nuclear revival" still remains an open option. But it will only remain so if the public authorities provide the sector with stability and predictability. Without those, an industry with fixed costs and high investments such as nuclear energy cannot remain competitive over the long-term.

1.4. Conclusion

At the end of this sociohistorical account on the trajectory of the French nuclear energy ecosystem, let us dwell on its properties – singular or not – and its specifically associated challenges, before re-examining the link, both strong and ambiguous, between the "technical" and the "political" components characterizing the French civil nuclear industry.

This section began with and discussed an important guiding question: Is the French case exceptional? In view of the situations observed in other nuclear-powered countries, we can respond positively to this, and outline the contours of the exceptional character of the French electronuclear sector. Its origin lies in the substantial share of electricity of nuclear origin, posing in a meaningful way the question of the dependence of France in relation to this mode of energy production. It also makes part of the technopolitical regime on which the French nuclear ecosystem is based. The latter is remarkable for at least two reasons: its small number of players (and their respective weight in decisions relating to energy issues) and the fact that for several decades, it has sought to maintain a clear boundary between the technical and the political. The operation of French power plants relies on long-term relationships between EDF and the regulator, formed by the Nuclear Safety Authority (NSA) and its technical support, the Institute for Radiation Protection and Nuclear Safety (*Institut de radioprotection et de sûreté nucléaire* [IRSN]), constituting an exceptionally stable "organized action system". This stability has contributed and continues to contribute to the robustness of the system in vigor, benefiting from strong feedback and operating and safety rules which have been tested and refined over time. But this stability can also work as a source of inertia and a relationships bureaucratization, with potentially harmful effects for decision-making, the ability to adapt and innovate in a largely uncertain environment.

In contrast to the remarkable stability of the system on which reactor operation is based, the design/construction activities of new reactors are characterized by an intermittent regime. These activities have the particularity of being both highly institutionalized and destabilized. Two long-term contradictory effects can be observed. The long history of reactor construction and operation generates high expectations from regulators, public authorities and the civil society: safety, reliability, industrial and economic performance requirements. But the discontinuous trajectory of design and construction activities over the last few decades, alternating peaks and troughs of activity, has also led to a disintegration of skills and the industrial fabric, as well as a destabilization of the organizations involved. The problems encountered by its flagship projects, EPR and ASTRID, attest to the profound difficulties these activities, and the organizations responsible for them, are confronted with: the phenomenon of organizational forgetfulness and loss of competencies, the instability of organizations in charge of design activities, inter-organization tensions due to contested roles and scopes of action, the inability to break away from the legacies of the past and to innovate, etc.

These destabilization phenomena, which largely preceded the Fukushima accident, are above all endogenous to the nuclear ecosystem and part of a long-term problem. They are set in this ambiguous framework of weakened organizations subject to ever-increasing demands. The major difficulties these projects are facing can be partly explained by the fact of having to resolve these long-term contradictions in the very short-term. The inability of stakeholders to recognize and then manage the intermittency problem in design/construction activities contributes to the current pitfalls: projects of an uncertain or even hybrid status (Tillement and Garcias 2021), between the innovation and renewal of the old, role conflicts between historical actors, not to mention the divestiture of some. Depriving the players of the possibility of constructing clear and shared goals and frameworks, of assessing performances and producing trade-offs, this context exacerbates the tensions and balance of power between operators, designers and regulators. Project stagnation also accounts for the delegitimization of the nuclear sector. Without foreshadowing the future of the nuclear industry, the ability of manufacturers, regulators and the public authorities to recognize the major variations in activity levels in the design of reactors as well as to understand their long-term consequences seems a prerequisite for the joint construction of approaches.

The dawn of the 2020s is a time for making choices regarding the French nuclear ecosystem and France as a country. The State is called upon to question itself and to quickly take a stand on major issues. Can France do without nuclear energy, and if so, what other alternatives are available? By what deadline could we reasonably envisage a nuclear power phase-out, if this is democratically decided? Conversely, does France wish to pursue nuclear energy, and if so, in what form? By maintaining global leadership in this area, what would imply maintaining and renewing skills, and, subsequently, the rapid launch of new mass production programs? Or, failing that, what about keeping nuclear in the electricity mix, with weakened competencies, recreating a technological dependence from foreign powers (South Korea, China, Russia), as was the case during the 1970s, before France chose to emancipate from dependence on American companies? This second option would necessarily entail renouncing a part of France's energy independence.

Since nuclear power is a capital-intensive industry, assuming long-term strategies and investments, these decisions should, in any case, be made quickly. An apparent non-choice (or a permanently postponed decision) ends up being a very strong choice: most likely, it could lead to France's loss of leadership in the nuclear industry to the benefit of the powers investing in the field, essentially China and Russia. For nuclear power, as for other energy technologies, the present choices undeniably involve a central geostrategic dimension.

1.5. References

Boiteux, M. (2008). Maîtriser les noyaux et les pépins. *Le jaune et le rouge – Revue mensuelle de la Société amicale des anciens élèves de l'École polytechnique*, 637.

Bréchet, Y. (2019). L'arrêt du programme ASTRID : une étude de cas de disparition de l'État stratège. *Revue progressistes. Actualités*, 25.

Danby, G. (1996). *Science and Environment Section*. House of Commons Library, London.

Deffrennes, M. and Iracane, D. (2020). Le nucléaire civil, enjeux et débats. Evolution du cadre d'activité de la filière de l'électronucléaire au niveau mondial : de 1950 à aujourd'hui. *Responsabilité & Environnement*, 97.

European Union (2010). Traité EURATOM. Version consolidée. Report, Office des publications de l'Union européenne.

European Union (2012). Traité sur le fonctionnement de l'Union européenne. *Journal officiel de l'Union européenne*.

Finon, D. and Staropoli, C. (2001). Institutional and technological co-evolution in the French electronuclear industry. *Industry and Innovation*, 8(2), 179–199.

Garcias, F. (2014). Apprentissage, désapprentissage et réapprentissage organisationnels : le cas d'une activité d'ingénierie de grands projets complexes. PhD Thesis, Mines Paris Tech, Paris.

Grubler, A. (2010). The costs of the French nuclear scale-up: A case of negative learning by doing. *Energy Policy*, 38(9), 5174–5188.

Hecht, G. (2004). *Le Rayonnement de la France : énergie nucléaire et identité nationale après la Seconde Guerre mondiale*. Le Seuil, Paris. [Hecht, G. (2000). *The Radiance of France*. MIT Press, Massachusetts.]

HM Government (2018). Industrial strategy: Nuclear sector deal [Online]. Available at: https://www.gov.uk/government/publications/nuclear-sector-deal/nuclear-sector-deal [Accessed June 27, 2022].

HM Government (2020a). Energy white paper: Powering our net zero future [Online]. Available at: https://assets.publishing.service.gov.uk/government/uploads/system/uploads/attachment_data/file/945899/201216_BEIS_EWP_Command_Paper_Accessible.pdf [Accessed December 2022].

HM Government (2020b). The ten point plan for a green industrial revolution [Online]. Available at: https://assets.publishing.service.gov.uk/government/uploads/system/uploads/attachment_data/file/936567/10_POINT_PLAN_BOOKLET.pdf [Accessed November 2022].

Nuttall, W. (2004). *Nuclear Renaissance: Technologies and Policies for the Future of Nuclear Power*. Routledge, Abingdon.

OECD/NEA (2012). Nuclear education and training: From concern to capability [Online]. Available at: https://www.oecd-nea.org/upload/docs/application/pdf/2019-12/6979-nuclear-education.pdf.

Sénat (2000). Rapport d'information du sénat n° 320. Annexe au procès-verbal de la séance du 2 mai 2000. Report, Sénat.

Taccoen, L. (2003). *Le pari nucléaire français : histoire politique des décisions cruciales*. L'Harmattan, Paris.

Tillement, S. and Garcias, F. (2021). ASTRID, back to the future: Bridging scales in the development of nuclear infrastructures. *Nuclear Technology*, 207(9), 1291–1311.

Tillement, S., Journé, B., Thiollière, N., Mouginot, B. (2014). Les échelles de temps des scénarios électronucléaires. In *Le nucléaire au prisme du temps*, Bretesché, S. and Grambow, B. (eds). Presse des Mines, Paris.

Torres, F. (2016). Le système nucléaire français des années 1950 à nos jours, acteurs et structures. Une mise en perspective. *La revue de l'énergie*, 634, November–December.

UNSCEAR (2020). Scientific Annex B. Levels and effects of radiation exposure due to the accident at the Fukushima Daiichi Nuclear Power Station: Implications of information published since the UNSCEAR 2013 Report. Report, UNSCEAR.

2

Nuclear Costs: Methodological Aspects

Jean-Guy DEVEZEAUX DE LAVERGNE[1] and Nicolas THIOLLIÈRE[2]

[1] *Formerly I-tésé (CEA), Gif-sur-Yvette, France*

[2] *Subatech, IMT Atlantique, Nantes, France*

2.1. Introduction

The issue of energy costs has been the subject of numerous articles in the dedicated economic literature for decades. In the context of the fight against climate change, the main question concerns the future role of various electric energy sources, including nuclear power, and the investment choices they entail. But questions have also significantly evolved over the past 20 years and have diversified. The demand for studies related to energy transition has grown tremendously, particularly on the part of public authorities establishing energy transition plans.

The major issue in the last decade has been the development of renewable energies, namely variable ones. This has upset the historical framework regarding the management of electricity markets and the economic analysis of optimal nuclear fleets. The notion of system cost was theorized during the 2010s (OECD/NEA 2012 2019; Hirth et al. 2015). The question of greenhouse gas emissions from electricity production means, which is only part of the broader question of electricity production externalities, is in itself input data for the analysis of nuclear power plants.

Nuclear Economy 1,
coordinated by Jacques PERCEBOIS and Nicolas THIOLLIÈRE.

From a methodological perspective, works acknowledging the long-term or increasing uncertainties have marked recent times, together with the formalization and evaluation of production costs.

Compared to previous works and reports on the economy of nuclear energy (Bertel and Naudet 2004; D'Haeseleer 2013; Rothwell 2015), the questions posed are now significantly renewed (Devezeaux and Berthélémy 2020). The focus has shifted toward reactor lifetime extension, investment dynamics in interconnected production plants under strong technological change (possibly including the production of heat or others), or the preparation for the deep decarbonization of our economies, which is urgent before reaching the mid-century. Paradoxically, an important development – the arrival of the historic fleet to its maturity – has resulted in an increasing interest in the cost of existing nuclear energy.

2.2. The different notions of costs

Basically, the concept of cost comes down to calculating the monetary equivalent of the consumption of a good or the provision of a service, for a given actor, in a given context (technical, economic, political, regulatory, social, etc.). In this case, the good is electricity. This almost immaterial good is measured in energy, although its value also strongly depends on the moment of its production and consumption. The existence of a market aims to even up the value of this good and its production cost. In a perfect market, the price reflects such balance. But the electricity market is a complex one and involves a whole set of production and regulation technologies, a transmission network, etc. Time plays a major role in this market, since balance needs to be maintained at every moment. No nuclear electricity market could inform us about the production and provision costs only by looking at the price. Instead, there is a great deal of accounting information for appraising the different cost indicators (like a sum of all costs: salaries, the capital immobilized in power stations, intermediate consumption, etc.), calculated for each unit of energy produced in kWh or MWh. These costs are broken down in space (geography, perimeter and agents concerned) and time. Calculated today, they are generally used to assess the performance of past choices (have decisions delivered on their promises?) to shed light on pricing choices in a regulated economy scenario (how to bill the electricity produced by historic nuclear power?) and to clarify investment choices (how to build the electricity mixes of the future?).

2.2.1. *Nuclear costs and associated challenges*

Calculating the cost (in this case, the production cost) of nuclear-generated electricity is thus a more complex exercise than it seems. Most importantly, there is no single definition of the cost of a good or a service. The methodology used for calculating a cost primarily depends on the use that will be made of it (the question that will be asked). In broad terms, nuclear electricity comprises two main cost categories:

– Accounting costs – actually, production costs – which aim to clarify the decisions of economic agents (the producing company, in the first place) by summing up the various expenditure items required for producing electricity. It should be noted that, even in this apparently simple case, the accounting system must be able to take into account capital expenditure (fixed assets) history and the constitution of a future expenditure "stock" (decommissioning, waste management, etc.).

– Economic costs, which aim to provide information on the impact the production of nuclear electricity generates in terms of wealth. This concept assumes that markets can translate (even with some corrections) the balances leading to the social optimum. This includes the scarcity of goods, the environment and health. The economic cost is the combination of losses of any goods that have a value attached to them by any one individual or group. In particular, these economic costs are used for guiding public choices. Even if they are initially calculated from the same information as accounting costs, they supplement them with further information and special methodological approaches apply.

The calculation of a cost also depends on the field it intends to cover, which leads to structuring the field of analysis. The main dimensions of this field are the agents (which are the economic agents concerned?), the temporal dimension (from the very short to the very long-term) and the spatial dimension or even the perimeter (what are the effects to be taken into account, which mesh and where?).

2.2.2. *Defining the actors concerned by an economic question*

Cost calculation will not be the same depending on whether one is interested in one economic agent or another. In terms of production costs, the first actor concerned is the operator. But the production cost is affected by numerous rules, factors and provisions that generate flows with other actors (or economic agents): employee remunerations, taxes and subsidies involving local communities and the

State, for example. Other impacts, whether monetary or not, also result from investments or the operation of one reactor alone or multiple nuclear power plants. Modifications to the functioning of electrical systems (balancing, network needs, etc.) are systematically required, depending on the production means chosen. They primarily concern another actor: the network manager. Electricity consumers are also directly affected by prices and tariffs. Local communities are affected by the economic activity, transport and taxes. The State is a major actor, and its choices are notably guided by the calculation of the project's value from the society's viewpoint (the nation). Greenhouse gas emissions concern all of humanity. According to these arguments, it is desirable to determine the full social cost of nuclear energy and other production modes.

Measuring the impact of a reactor's operation can be differentiated for several agents. This plays on the perimeters or depth of the temporal analysis, but not only that. Calculation methods also depend on the agents involved. An important nuance concerns the choice of discount rates. As private rates (relating to companies) are generally higher than public rates (relating to the evaluation of the costs for society), this may result in significant differences across costs of a similar nature, though calculated for different actors.

2.2.3. *Organizing cost concepts following a temporal logic*

Hence, the big questions can be classified according to a temporal logic, which will be completed in the following section.

Nuclear costs of older generations of reactors, which have already been shut down: these costs are largely due. Few problems are associated with them, apart from the search for an economic assessment of past choices or taking into account future expenses (decommissioning and waste not yet disposed of). For the record, these are as follows:

– Existing nuclear production costs: these costs are largely due (initial investment). Several concepts coexist. From the producer's point of view, there are the total costs, which integrate the whole economic cycle, and the cash costs (short-term cash flows disbursed). The associated problems and use of these costs are maintaining reactors in operation and calculating the production cost price in order to set the price by regulation. These costs can be determined from the electricity supplier's point of view or from a broader perspective. They can be calculated for a representative year or be related to total expenditure (and production) over a long period of time. Often, they only focus on future years.

– Costs of the nuclear facilities yet to be built: here, all costs refer to the future. They are essentially used for deciding whether the new investment will be made or not. They are generally discounted (with different rates depending on the actors). In addition to the electricity generator, these calculations can and should take further impacts into account, in particular on the electrical systems. Since investment choices influence the public domain, and must acknowledge many other aspects, such as environmental and health externalities, the impacts on national sovereignty, the effects on employment or the trade balance (in the event of imbalances in these items), etc.

– There is also a "hybrid" calculation which combines the two approaches: it assesses the decisions for *lifetime extensions,* when significant safety upgrade costs are incurred.

In addition to the above-mentioned economic considerations, which focus on the generations of reactors, it is necessary to choose an adequate time frame. Over short periods of time, almost all costs are fixed: the *short-term variable cost* of the unit is key for making decisions in terms of market prices (above or below) and whether or not to engage the production means.

On the other hand, an annual basis informs about the equipment's profitability or the justification of a selling price, in particular regarding the concept of *total cost.* For longer time spans (the entire reactor project), it is possible to focus on the overall value generated by the equipment or a thorough program.

2.2.4. *Specifying the scope of cost analyses*

As it is becoming clearer at this stage, choosing the scope of the notion of cost is crucial. Some previous studies (Hirth et al. 2015; OECD/NEA 2015; Samadi 2016) provide a rather complete overview, briefly presented in Figure 2.1. As a general rule, the more global the actor concerned, the larger the scope of the analysis: from the electricity supplier (microeconomic analysis), including the network operator (who must take systems costs into account), to the State (which synthesizes a broad field and often a larger dynamic). Several costs can be thus identified.

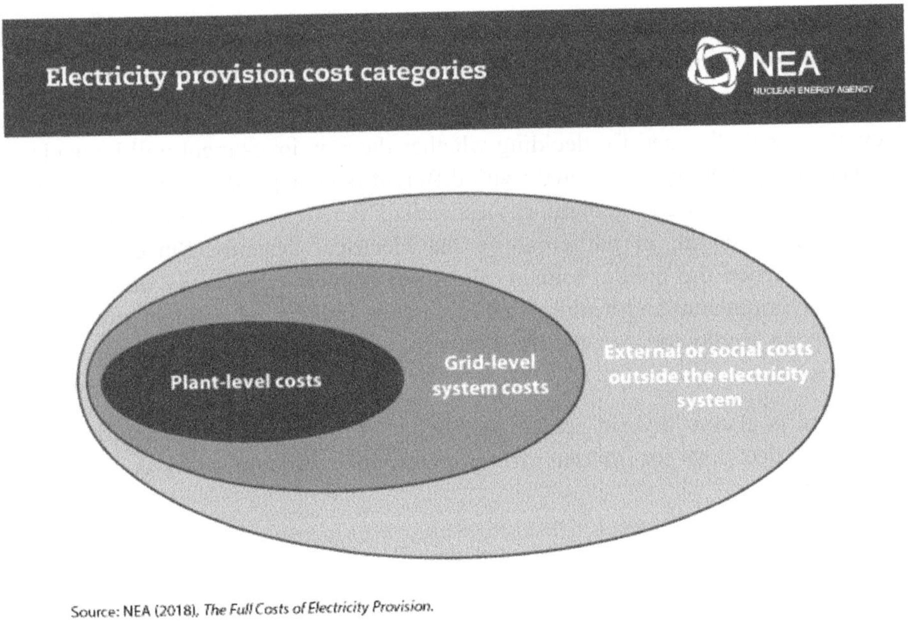

Figure 2.1. *The various scopes of cost analysis according to OECD/NEA (2015). For a color version of this figure, see www.iste.co.uk/percebois/nuclear1.zip*

2.2.4.1. *Production cost "at the power station terminal" and its main components*

The main components are as follows:

– Investment costs, including engineering costs, "owner costs[1]" and construction costs, in the ample sense. A special case refers to operating life extension, which is comparable to a new investment during operation.

– The cost of capital (e.g. interests on loans; this cost also generally includes the expected return on the project, depending on the risks incurred).

– Operating and maintenance costs (which mainly include manpower costs, intermediate consumption excluding fuel, such as subcontracting and taxes).

– Fuel and associated waste management costs.

– Decommissioning costs (and those associated with waste management), which can also be considered as capital costs (and therefore included as investment costs).

1 These costs include administrative expenses, taxation schemes, regulatory costs (e.g. public inquiry), environmental monitoring, communication, etc.

Construction costs deserve a closer examination. They include administrative expenses, taxation schemes, regulatory costs (e.g. public inquiry), environmental monitoring, communication, etc. A description is available in D'Haeseleer (2013) and OECD/NEA (2020b). These costs can be broken down into site costs (purchase of land), site preparation, strictly speaking construction costs, start-up and pre-operation costs, transfer costs to the operator, etc. In strict terms, construction costs are the heaviest item and can be broken down into civil engineering items, nuclear island equipment (reactor and first core) and "conventional" island equipment (turbines, alternator, cooling, etc.), bearing in mind that sometimes the electrical equipment is discriminated. Each cost can also be broken down by type of expense (discriminating salaries, etc.). This nomenclature has the advantage of being consistent with the corporate accounts (income statement, etc.). Finally, it is possible to draw a distinction between direct costs (material, equipment and manpower) and indirect costs (project management, engineering, quality assurance, etc.). This logic is common in engineering because most of the *design-to-cost* is related to direct costs, while the rest of the costs depend on them more or less proportionally (insurance, for example), or are exogenous (public inquiry costs, for instance). In quotes (before construction or ex ante), all these items are subject to contingencies (typically 10–20%).

It should be noted that the disbursement dates corresponding to these five cost categories are not the same. The first are spent before operation, the second ones can extend across the project's entire life (depending on funding arrangements), the third are directly associated with the operating rate, whereas the last take ones place after decommissioning, sometimes several decades later. Acknowledging the time factor during the project's assessment is of the utmost importance. In this sense, interest can constitute a significant part of the costs. *Overnight costs* refer to investment costs which do not take interest into account[2].

The total production costs at the power station terminal are typically between €50 and €100/MWh for a reactor yet to be built. This notably depends on the country, technologies and discount rates applied. The production cost is generally slightly lower (subtract €10–30/MWh) for an existing reactor.

These notions will be developed in further detail in Chapter 3 (cost calculations), where they will be evaluated per component.

2 As if the reactor had been built "in one night".

2.2.4.2. *The costs of switching from a complete electrical system to a dynamic one (system costs)*

These costs acknowledge the complete operation of the transport and distribution network (in the short and the long-term). In addition to the previously discussed costs, there are the following ones, as proposed by OECD/NEA (2012, 2019):

– The costs of connecting production and network units (lines). These may vary, depending on the spatial distribution of the production and the demand (particularly during the peak).

– Costs related to the short-term supply-demand equilibrium which is modified by the existence of the equipment studied (positive or negative costs). These costs strongly depend on the more or less predictable character of the production units.

– So-called "profile" costs, resulting from the addition of a new unit to the existing fleet. They are related to the time profile of production units and their hourly availability (example of solar PV generators and day/night alternation, or nuclear energy during reloading).

These system costs can be significant and affect the competitiveness of the various electricity production means. They typically amount to tens of euros/MWh and are lower for nuclear than for other production means, because nuclear energy is controllable and generally available, which greatly limits balancing and *backup costs* (including storage), compared to variable energies. These costs are not externalities: they are incurred (and paid for as additional storage, lines, etc.) by the various actors on the network (but not necessarily by the one who generates them).

2.2.4.3. *Other cost components, outside the electricity system (essentially, external costs)*

By nature, these costs are numerous. In the past, they have given rise to evaluation attempts at the national or European level. The majority of them do not give rise to exchange on markets, and therefore constitute externalities (the monetary valuation of these costs does not result from the market and, if possible, must be determined by special methods). Some costs can also change status: an externality can become a real cost for operators due to regulations "internalizing" them, even if partly (taxes on pollutants or market emission rights). In that case, they must also appear on a new list of costs for the end product.

The detriments (or benefits) associated with the establishment of a production unit can be monetarily assessed: pollutant emission (including CO_2, nitrogen oxides, particles, solid waste, etc.) and their impact on ecosystems, biodiversity (Brook and Bradshaw 2014) and human health (morbidity, mortality), and the impact on the value of nearby housing production units, on the landscape, noise and even on the territories' reputation. In addition to these impacts during "normal operation", we can mention the risks of accidents and associated costs. The agents concerned are numerous and diverse.

Other consequences are detectable at the economic and social level. They can give rise to attempts at macroeconomic evaluations, such as the impact on unemployment, real household income or the trade balance. Finally, other approaches can assign values to power-generating equipment, such as scenario analysis. These will be discussed further.

A fairly complete list of these types of costs can be found in some previous work (Samadi 2016, 2017; OECD/NEA 2019). Their evaluation remains tainted by strong uncertainties to such an extent that these costs are too often omitted in the studies of nuclear power plants. In total, however, they can amount to tens of euros/MWh or even more. Their role is crucial and no one should ignore them.

From the information listed above, various cost indicators can be drawn. But in order to assess the performance of a particular reactor or fleet, the question also comes down to choosing the right calculation methods. These are summarized in section 2.2.5.

2.2.5. *Choosing the appropriate calculation methods*

The economic analyses of nuclear energy can be deployed on three axes: the economic agent concerned (it is according to their "point of view" that analyses will be carried out), scope and time frame. The choice of a cost calculation method involves positioning oneself on the three axes, depending on the question asked and opting for a cost concept, with the nomenclature and the relevant calculation methods. This concept can be accounting-oriented (with a generally "microeconomic" perspective, centered on the electricity producer) or of an economic nature (rather with a social questioning, often relating to public decision-making).

In a restrictive way[3], the main economic methods for analyzing nuclear costs are as follows:

---

3 In fact, almost all assessment methods can be applied to nuclear energy, which is hardly specific. In particular, the methods for acknowledging uncertainty and risks constitute a vast field which can only be harnessed by a few tools (essentially, discounted cash flow analysis). We will also remain within the classic field of cost evaluations, making little mention of multi-criteria approaches (see, for example, IAEA 2019).

– **Analytical cost calculation**: this is essentially an accounting method. It is based on business accounting, breaking down all costs by product line (e.g. sale of electricity produced by a given reactor). It basically aims to calculate an annual cost price. The apparent simplicity of summable analytical costs actually hides a lot of know-how and method choice. Most of this knowledge relates to the allocation of cost lines to each of the product lines. These questions will not be addressed here, because they are specific to the actors implementing them. The very nature of this approach is to set the selling prices (or conversely to calculate the margins if the prices are exogenous) by product line. This approach often results in the calculation of *cash* costs: these are the analytical costs which give rise to effective disbursements (unlike, for example, depreciation or other provisions allocations).

– **Discounted costs and consideration of the time frame**: as cost studies generally span long periods, economic calculation must aggregate flows from different dates. This type of method is generally applied to investment choices, with a view to evaluating a future project. In this sense, this approach is also accounting and analytical in nature[4]. An essential portion of economic calculation for nuclear energy is dedicated to the evaluation of discounted costs. This will be described in detail in section 2.5.

– **Scenario methods**: these describe the evolution of the electrical system as a whole. They integrate the reactors' production costs (often with the need for annual breakdown) into a broader economic analysis. In short, these methods (often based on an updated approach) make it possible to compare various scenarios. Although it is not the subject of the present chapter, the probabilistic scenario approach (in particular, as a support means for real options calculations) can make it possible to assign a specific value to the future existence of a technology via its current implementation[5].

– **Macroeconomic modeling**: this is a set of tools for comparing scenarios via macroaggregates (GDP, employment, etc.), often following a de facto multi-criteria approach. It provides impact assessment, for example, when comparing nuclear

4 However, data collection is of a different nature (since the goal here is not to allocate actual costs to product lines, but to construct these costs with the analyses of future costs), and the main difficulties essentially concern anticipation errors.

5 This is particularly the case when a country risks losing control of a technology, unless it develops it. We will consult the study by Taverdet-Popiolek and Shoai Tehrani (2015) on fast reactors in France, which gave an option value of several billion euros to the construction of the Astrid international prototype. A similar question arose in 2021, in relation to the launch – or not – of an EPR nuclear program. In a similar spirit, the article by Gollier and Devezeaux de Lavergne (2002) evaluates the value of the reversibility function for a radioactive waste repository.

energy/other technology, but no single cost indicator. Such tools can be used for calibrating essentially macroeconomic corrections (in this case, imbalances on certain markets), so as to include them in microeconomic calculations, by correcting the latter (via corrective factors for certain positions). The results of these tools can feed evaluations applied to one reactor, insofar as the analysis claims a significant scope beyond the net present value (NPV[6]) (e.g. a differential of jobs following a strategy more or less based on nuclear energy can be summarized as jobs/reactor).

– **Non-"strictly" economic approaches** (also relying other human and social sciences): other human sciences, apart from economics, have long been involved in energy and nuclear issues, and their work can also be taken into account for decisions. Many of these works relate to political decision-making (often in terms of acceptability or – better still – socio-technical feasibility) or to risk threats to people and property, particularly in terms of perception. While this work has a large scope, it only rarely gives rise to specific approaches in terms of costs.

– **Multi-criteria methods**: apart from the assessment by cost (economic evaluation and other tools and concepts), methods can take into account quantifiable parameters, but choose not to aggregate them via a monetary assessment (e.g. the impact on biodiversity), preferring multi-criteria approaches. Thus, the NEEDS EU project (2009), for example, proposes a multi-criteria method focusing on nuclear energy. It is also possible to take into account discontinuous indicators (e.g. "favorable" and "unfavorable" appraisals), with dedicated multi-criteria methods. These approaches can integrate almost all the possible assessment criteria, but results are closely linked to the group of evaluators (*focus group*) via the "weight" of each indicator.

The combination of these methods, from a social perspective, is only applicable to a certain extent by aggregating relevant and reasonably quantifiable cost components. The resulting index is positioned on the largest possible scope of Figure 2.1: following the (OECD/NEA 2020a) nomenclature, it shall be called the *full* (social) *cost*. This concept, albeit difficult to implement, is the best tool for comparing electricity production systems from a public perspective.

2.2.6. *Data*

Nuclear stakeholders have often been criticized for a lack of transparency in this area. However, it should be kept in mind that, almost since the origin of nuclear electricity, economic calculations have been made public, in terms of production equipment choice. This is particularly the case of France where the PEON

6 The concept of NPV is detailed later in this section.

committee issued numerous reports from the very beginning[7]. Many countries have followed similar approaches (United States, Germany, UK, etc.), both in terms of electricity production or end-of-cycle choices. Internationally, the OECD has been publishing studies for several decades. These studies often detail highly accurate data in terms of investment costs. In general, the original sources are the electrical companies, data being collected by the public authorities. Currently, in France, data are mainly disseminated in ministerial reports and databases, those issued by the Parliament (OPECST[8], in particular), the Court of Auditors and RTE (*Réseau de Transport d'Électricité*). Companies also play a very important role via annual reports and numerous communications, some of which are specifically governed by the law, in special, the provisions and dedicated funds for the back-end fuel cycle, waste and decommissioning, in accordance with Article 20 of the law of June 28, 2006.

The nature of data is multiple-faceted: it depends both on the economic aggregate assessed (on its nature, the data organization plan and its scope) and on the use of such data. These uses can be divided into three major groups: project cost assessment, business activity monitoring and compliance with current cost-analysis approaches. Data relating to investment costs are evaluated by the engineers and their project managers (ex ante and during construction) and progressively translated – corrected – in the companies' accounts (as fixed assets, after start-up, following a different nomenclature and time frame). Similarly, future expenses are assessed by the engineering departments (according to their own nomenclatures) and translated into the social accounts (usually with less detail and processing discounts). The operating cost data (operation and fuel) are generally annual charges and the amounts recorded in social accounting remain close to their analytical equivalents, discounted or not (unless there are long-term expenses).

2.3. The discounted cost of nuclear electricity

This section aims to introduce and clarify the concept of levelized cost of electricity (LCOE) concept and the discounting principle associated with it. Let us specify the stakes of this calculation within the framework of a decision-making process, be it public or private. The methods for calculating the discount rate will also be discussed in order to give the reader the necessary keys for understanding the debate. The literature offers numerous sources of theoretical and applied

7 For the production of nuclear-generated electricity, which studied the ex ante profitability of electrical investments in France in the 1960s and 1970s.

8 Parliamentary Office for the Evaluation of Scientific and Technological Choices, bringing together the Assembly and the Senate.

information. Here, the focus is on the many reference reports and publications issued since the 2000s.

2.3.1. *Roles and challenges of economic calculation*

2.3.1.1. *General aspects*

"Economic calculation" refers to the set of methods making it possible to evaluate the collective opportunity of a public project on the basis of a calculation synthesizing its advantages and costs. In certain cases, economic calculation methods (and this is the case of the LCOE) can also apply to other economic agents (in which case one speaks of private economic calculation), but its theorization is mainly centered on the State's role. Regardless of whether the question involves a choice of investment, pricing, a taxing scheme, etc., economic calculation is often based on inter-temporal arbitration. Indeed, costs and benefits (economic or not) are usually out of sync and can span relatively long periods of time. In this context, "inter-temporal choices" are omnipresent and have been extensively discussed in the specialized literature[9]. Their history is transcribed in Frederick et al. (2002).

Economic calculation is a tool particularly used by the public authorities. One of the strategic roles of the State is to "arbitrate between different actions having a present and future impact on the well-being of the population" (Gollier et al. 2011). Many fields of action are concerned. To mention a few, these include the fight against greenhouse gases or the management of nuclear waste, for example. In the context of a public action involving limited resources, the State must arbitrate on the basis of a hierarchy weighing various future projects. By now, it should be clear that the calculation of a LCOE is done for a particular case: the study of a new electricity production project.

Economic calculation is one of the support tools for the decision-maker, whether political or industrial (*ibid.*). Historically, the importance of economic calculation has been widely stressed by the General Planning Commission, which became the Strategic Analysis Center in 2006 and then the General Commission for Strategy and Prospecting (France Strategy) in 2013. This calculation is based on a discount rate which assigns value to time and provides a correspondence rule for harmonizing desynchronized monetary flows. The choice of a discount rate is all the more important the longer the time horizon covered by the project (Schubert 2006).

9 In this debate, we can highlight the key roles of the publications by Fisher (1930), Samuelson (1937) and Koopmans (1960).

Any reflection on the discount rate must refer back to the fundamentals of economic calculation, because it constitutes one of its pillars. The dissemination of calculation methodologies could shed light on the debate between the different stakeholders (Lebègue et al. 2005). It should be noted that this last point, which is particularly relevant, is emphasized in the numerous reports mentioned. In Gollier et al. (2011), it is pointed out that "the interest of economic calculation is not to summarize the social value of a collective decision in one indicator or other, but rather – by forcing agents to produce them – to challenge them, even to enrich them, to reason the debates in a shared frame of reference".

2.3.1.2. *Specificity of the energy sector*

In Gollier et al. (2011), the socioeconomic evaluation in the field of energy is discussed. The authors particularly stress that even if the energy sector relies heavily on socioeconomic calculation, this practice is in sharp decline. However, the particular context of the energy transition requires decisions which, over time, will amount to several hundreds of billions of euros. The energy context is very complex and uncertain, with distant horizons. Economic calculation seeks to provide formalized coherence, clarifying the evaluation of the different possible trajectories. In this particular context, the determining role of the State is of great importance. The latter counts with action levers relating to the energy supply and demand, as well as a legislative framework.

The nuclear energy generation sector uses economic calculation methods for at least two reasons. First of all, the time horizon of a nuclear investment project is potentially very long. It extends from several decades for a nuclear reactor up to a hundred years in the case of nuclear waste storage or a disposal center. The exploitation of nuclear energy is usually backed by State support, thus securing investments and guaranteeing the interests and/or missions of the stakeholders (operator, safety authority, consumers, etc.). As the State intervenes on issues relating to energy and in particular to nuclear energy (taxation scheme, market organization, etc.), the question arises as to the optimal allocation of public money. It is worth recalling the interest of economic calculation applied to public decision-making.

2.3.1.3. *Public and private economic calculations*

The report by Gollier et al. (2011) addresses the question of the degree of uncertainty inherent in any human decision. The decision involves taking an inevitable risk. Economic calculation is one of the decision support tools used in both the public and private sectors.

Private enterprises use economic calculation as a tool for choosing an investment strategy based on the optimization of economic profit. Evidently, risk analysis is also a key element of the decision-making process. This last element can even replace the expected internal rate of return if the project's risk is too high.

Economic calculation as implemented by the public sphere has a different purpose. It aims to "objectivize the social utility of public expenditure by leveling all the non-market effects of a decision or an investment under the same unit of measurement" (Gollier et al. 2011). In Lebegue et al. (2005), it is stipulated that the public decision-maker must "focus on the socio-economic profitability which reflects the interest of a project for the community as a whole (users, public authorities, taxpayers, concessionary companies, communities territories, local residents, etc.) by taking into account elements other than cash flows alone: security, pollution and nuisances, scarcity of certain natural assets such as energy resources or raw materials, expected scientific and technological progress, the effect emissions and climate change, consumer and user behavior, economic and social risks, etc."

2.3.1.4. *Limitations of economic calculation applied to electricity*

Economic calculation is subject to many legitimate criticisms. First of all, the practice of economic calculation involves integrating all the costs and advantages related to the project, be them monetary or not. However, these flows can be highly heterogeneous and need to be converted into monetary values in order to be correctly aggregated (Lebègue et al. 2005). This work will discuss the case of externalities applied to the production of electricity. Economists will try to measure such externalities in monetary terms so as to include them in the calculation of LCOEs (as well as in most cost calculations, from the community's viewpoint). But, within the framework of public economic calculation, the benefits are not necessarily monetary (operation lifetime increase, for example). It is then necessary to resort to specific techniques to encrypt all the flows and thus be able to compare them.

In the case of the LCOE, a major limitation of the concept is related to the fact that calculations are generally limited to the scope of the source of electricity production, without taking into account the operation of this source within the electrical system. However, every piece of equipment reacts in a specific way to the control of the supply-demand balance and acts on the electrical system for costs (and rarely benefits) to ensue. The various system costs are presented in Chapter 3. These can be very important and the sole examination of the LCOE "at the power station terminal" is no longer sufficient to qualify the competitiveness of this equipment, mainly due to the massive arrival of renewable energies.

Another limitation lies in the choice of the discount rate. In theory, the public rate must be unique and must be "uniform to all public investment projects". However, it is difficult to guarantee the harmonization of practices. This is particularly remarkable in the field of energy prospecting. The economic evaluations of energy trajectories or energy mixes at a given time are highly dependent on the rates chosen. Without a standardization of practices, it is very difficult to compare studies.

The report by Gollier et al. (2011) also discusses the difficulties encountered by economic calculation. The problems of inter-personal and/or heterogeneous comparisons, as well as temporal arbitration, are stressed. Finally, let us recall that the management of risk and uncertainties constitutes the most difficult challenge in the context of economic calculation.

Despite these constraining limitations, the application of discounted cash flow analysis to economic calculation and discounting appear as "tools that can be summoned by all the stakeholders" to provide an "explicit framework for reflection", valuing and weighing the benefits and costs of a project (Lebègue et al. 2005). In this sense, economic calculation is effectively a decision-making tool. While the aforementioned reports recognize the limitations of the available tools, they also emphasize their indispensable character for "feeding and enriching the debate and for preparing the arbitration, choice and essence of public governance". The contribution of economic calculation to fields involving a temporal horizon over several generations is definitely clear. Once again, the control of greenhouse gas emissions or the sustainable management of nuclear waste are unmistakable examples in this sense.

We also mentioned earlier in this chapter other methods which make it possible to address the impact of large projects (macroeconomic models, other human sciences), relying on different approaches.

2.3.2. *Key preliminary concepts for calculating the LCOE*

The presentation of the standard methodology for economic calculation is based on concepts, which must be mastered beforehand. These will be introduced in the following section.

2.3.2.1. *The basics of discounted cash flow analysis*

Discounted cash flow analysis is a technique used by economists for comparing the immediate effects of an economic action with the ones that will take place in the future (Portney and Weyant 1999). Inter-temporal trade-offs relating to the notion of discount aim to balance the "sacrifice" to be made today against the benefits expected in the future.

Discounted cash flow analysis is based on a relatively simple principle: "one euro now" is usually preferred to "one euro in the future" (Schubert 2006). In fact, it is always possible to invest this euro today in order to obtain a higher nominal value in the future. For outgoing monetary flows, the principle is symmetrical. Economic agents may prefer to save a euro, rather than spend it today, in the hope of generating more value in the future from this initial euro and consuming more over time. Conversely, Fery (2010) justifies the "preference for the present" as a consequence of risk aversion in a context of future uncertainty. To illustrate the concept, he quotes the famous maxim "one bird in the hand is worth two in the bush".

The concept of discount rate will be developed in the following section. At this stage, it should be kept in mind that a cash flow executed at a certain date t can be "updated" to a current date via the application of a discount factor equal to $^1/_{(1+\tau)^t}$ for discrete flows and e^{-Rt} for continuous flows. More generally, we call F_1 the flow on date T_1 and F_2 the one on date T_2. The equivalence between these flows is ensured by a discount factor such that:

$$F_1 = \frac{F_2}{(1+\tau)^{\{T_2-T_1\}}} \quad [2.1]$$

2.3.2.2. *Net present value*

It is important to grasp the concept of NPV to understand the principle of economic calculation. The NPV makes it possible to quantify the social utility of a project or an investment compared to a reference situation. The year preceding the commissioning of the evaluated project is called t_0. Its formal expression is written as (Lebègue et al. 2005):

$$\text{NPV} = -(I - I_e) + \sum_{t=1}^{T} \frac{F(t_0+t)}{(1+\tau)^t} - \sum_{t=1}^{T} \frac{\Delta I(t_0+t)}{(1+\tau)^t} + \frac{R}{(1+\tau)^T} \quad [2.2]$$

The NPV is calculated in constant currency, meaning that future flows will be expressed in the currency of the reference year (in this case, t_0). In this equation, I represents the discounted cost of the investment if the project is carried out over several time periods. I_e is the sum of the eluded investments: the investments of the reference case rendered unnecessary by the present investment. $F(t)$ represents the net cash flow, revenue minus expenses at date t. $\Delta I(t)$ is the cost of the major maintenance or upgrade investment at date t. Finally, R represents the residual value of the investment at the end of the time period for which the calculation is made. This parameter can be negative if site restoration costs are required.

Conventionally, the year 0 is often defined as the one preceding the facility's commissioning. A positive NPV indicates that the project is profitable. In order to illustrate this magnitude, let us propose a fictitious project defined by expenses and revenues, represented by the top curve shown in Figure 2.2. Construction begins at $t = -15$ until $t = -1$ and then follows a Gaussian profile. The operating period ranges from $t = 0$ to $t = 50$. Production costs increase as the facility wears out. Revenues are considered constant over the period. Finally, decommissioning expenses range from $t = 50$ to $t = 70$ and are characterized by a constant annual flow.

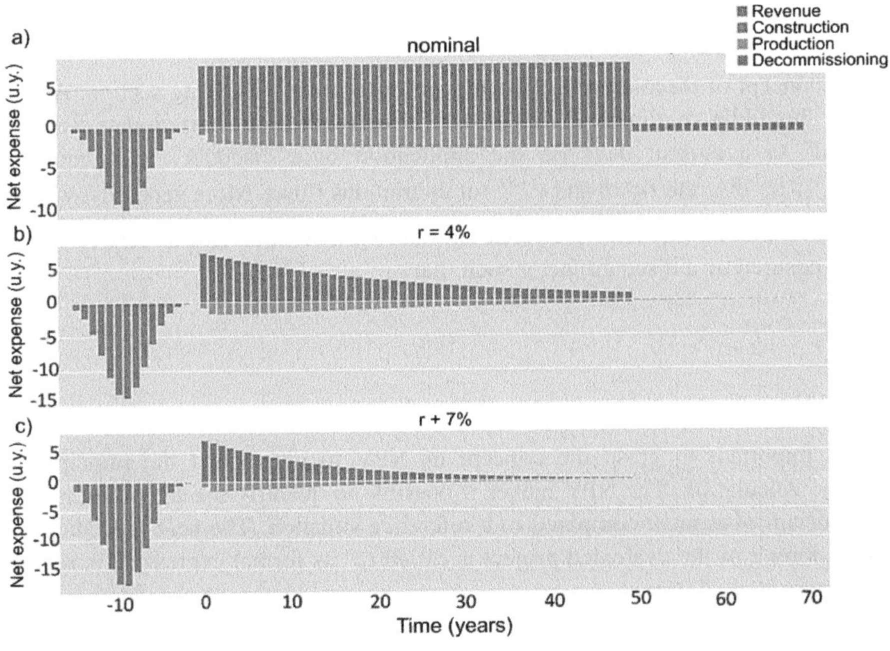

Figure 2.2. *Historic of expenses (negative monetary unit) and revenues (positive monetary unit) in arbitrary monetary unit for a fictitious project. (a) A zero discount rate; (b and c) produced at 4% and 7% rates. For a color version of this figure, see www.iste.co.uk/percebois/nuclear1.zip*

Then, a discount function (equation [2.1]) will be applied on each flow. Two discount rates (4% and 7%) are chosen, with a time reference set at $t = 0$. We then obtain the updated expenditure profile shown in Figure 2.2 (middle curve and bottom curve). Note that the discount function tends to increase construction expenditure (negative time index) and to "squash" monetary flows from commissioning, especially as the time horizon considered is distant. It should also be noted that, for this example, decommissioning expenses have become negligible in NPV.

The NPV is calculated as the sum of the discounted annual flows shown in Figure 2.2. It is clear that the value of the discount rate is a key parameter for the value of NPV. For this fictitious example, the "nominal" NPV, that is to say, with a zero discount rate, is 164 currency units. It is worth 22 currency units for a 4% discount rate and –38 currency units for the 7% discounted cost. We find that the discount rate can induce a positive or negative NPV. The assessment of the discount rate is discussed in section 2.4.

2.3.3. *Calculation of the LCOE*

This section succinctly illustrates the concept of LCOE, as well as its various components.

2.3.3.1. *General aspects*

By definition, the LCOE is the constant price of electricity that equates discounted revenues and expenses along the project's entire economic life. In this formalism, for the investor, the discount rate applied represents the return on capital in the absence of risk and without taking into account market effects (OECD/NEA 2010). If the producer sells electricity at the LCOE level, the expected margin is zero. This has led to another common definition for LCOE, as the minimum constant selling price at which an investor must sell his production if he wants to cover all his production costs (D'Haeseleer 2013).

This chapter is limited to the study of LCOE for a nuclear reactor project. But in practice, LCOE can be calculated for any electricity production system. For this reason, it is necessary to clearly explain the calculation hypotheses if one wishes to compare figures.

2.3.3.2. *The expression of LCOE for a "reactor" calculation*

To determine the first analytical form of LCOE (and depending on the breakdown chosen for the expression of the various cost components), the following quantities have to be defined for the year *t*:

– $E(t)$: electricity generation;

- $\frac{1}{(1+r)^t}$: discount factor;
- $I(t)$: investment cost;
- $OM(t)$: operation and maintenance costs;
- $F(t)$: fuel cost;
- $C(t)$: carbon cost;
- $D(t)$: decommissioning cost.

P_e s also set as the selling price of electricity considered constant under the LCOE formalism. Then we have the following equality:

$$\sum_t P_e\, E(t)\frac{1}{(1+r)^t} = \sum_t[I(t) + OM(t) + F(t) + C(t) + D(t)]\frac{1}{(1+r)^t} \qquad [2.3]$$

The sum is calculated for the project's entire economic life. In this formalism, the selling price will be deemed constant (without taking markets[10] into account), in which case the price for reaching this equality – by definition the LCOE – will be:

$$\text{LCOE} = \frac{\sum_t[I(t)+OM(t)+F(t)+C(t)+D(t)]\frac{1}{(1+r)^t}}{\sum_t E(t)\frac{1}{(1+r)^t}} \qquad [2.4]$$

There is an important discussion in OECD/NEA (2010) that is related to the denominator of the LCOE expression. Everything seems to indicate that the energy might be discounted despite the fact that the energy produced at each period is a physical quantity whose value does not change over time. In this case, energy must be considered as a physical flow for generating a flow of economic revenues, and this flow should be discounted.

2.3.3.3. *The expression of LCOE for a "material flow" calculation*

Another analytical form is possible for LCOE. It is based on De Roo and Parsons's (2009, 2011) work. The proposed approach explicitly aims to take into account the back-end fuel cycle and the recovery of materials by rewriting the LCOE, relating these costs to 1 kg of initial fuel. Although this theoretical approach will be briefly described, this chapter will focus on the simpler open cycle (the UOX

10 Or rather, that the reactor is contemplated from the perspective of a "price maker", not a "price taker". This hypothesis becomes less strict as soon as we compare the LCOE at the actual valuation price of the energy produced (unknown at the time of the investment, hence the market risk).

fuel goes through the reactor and the spent fuel is not recycled, but constitutes waste).

This approach works with industrial processes related to a unit mass of enriched natural uranium oxide fuel in a light water cooled reactor (PWR-UOX). Cycle operations and fuel manufacturing stages, like energy production, are thus reduced to costs per unit of mass passing through the system. It is not necessary to make hypotheses about the design or technical limitations of plants. With this approach, one can simply couple economic with physical data.

The open cycle is based on reactors loaded with UOX fuel, without recycling spent fuel, sent to deep geological disposal after passage through the reactor. The LCOE for this cycle is conventionally defined by a breakdown in the following components:

- front-end cost elements;
- reactor costs;
- back-end cost elements.

The front-end costs of the cycle are broken down into the cost of natural uranium u_1, the cost of conversion, enrichment and fabrication of 1 kg of UOX fuel. The sum of these last three terms is called b_1. Terms u_1 and b_1 are summed and divided by the "discounted" quantity of energy produced by the kilogram of UOX over the duration of the complete cycle in the reactor. We then calculate the front-end cost f_1 as:

$$f_1 = u_1 + b_1 \qquad [2.5]$$

The cost of capital k_1 is deduced from the overnight cost per kW (C_{CO}) and the construction schedule. It is customary to include the decommissioning cost (C_D) in construction costs. This cost is reduced to the kW installed. For simplicity purposes, let us assume the project will be seen through in 1 year. We then have:

$$k_1 = \frac{\left(C_{CO} \sum_{t=CT-1}^{t=0} \frac{W_t}{(1+r)^t} + \frac{C_D}{(1+r)^{LT}}\right) P}{E} \qquad [2.6]$$

In this equation, W_t represents the fraction of the construction cost spent in the year t along the construction schedule CT. Terms P and E, respectively, represent the reactor's electrical power and the "discounted" effective energy produced over the project's economic life LT. Note that, in the methodology by De Roo and Parsons (2009), additional terms are taken into account (taxes, depreciation, etc.).

Operation and maintenance costs m_1 are also included in the calculation as follows:

$$m_1 = \sum_{t=1}^{LT} \frac{C_{F-OM}}{(1+r)^t} \times \frac{P}{E} + C_{V-OM} \qquad [2.7]$$

In this last expression, C_{F-OM} and C_{V-OM} represent the fixed (per unit of installed power) and variable operation and maintenance costs.

Finally, because of this method, it is possible to directly express the back-end costs depending on the quantity of fuel (here 1 kg). Then, the back-end cost of the cycle amounts to d_1. In the United States, a sum from the fixed cost C_S (per unit mass of initial fuel), a long-term dry storage ST and a variable cost C_d/kWh are reserved for geological storage, because in the United States the electricity supplier pays a fee per kWh produced so as to transfer responsibility for the storage of spent fuel to the government:

$$d_1 = \frac{C_S}{(1+r)^{CT+ST}} \frac{1}{E} + C_d \qquad [2.8]$$

We then calculate the discounted cost of electricity l_1 for the system defined as the sum of all the above-mentioned terms:

$$l_1 = f_1 + k_1 + m_1 + d_1 \qquad [2.9]$$

The term l_1 represents the LCOE calculated following the "material flow" approach. This method can also be used for defining a fictitious price for recycled materials and for calculating the LCOE relating to advanced fuel management, such as the plutonium recycling in a PWR (case of France) or an FNR (DeRoo and Parsons 2011).

2.4. Determining the discount rate

The previous sections have insisted on the particularly relevant role played by the discount rate in economic calculation. This section aims to present the different approaches to estimate a discount rate that can be used in economic calculation for the public sector. Theoretical approaches, as well as a classical model, will be presented.

2.4.1. *The three theoretical approaches*

This section relies on the study by Lebègue et al. (2005) and their classification of the different approaches to evaluate the public discount rate:

– pure time-preference theory;

– the marginal return to private investment;

– long-term interest rates.

In an unrealistic case with a perfectly functioning market, these three approaches would be equivalent.

2.4.1.1. *Pure time-preference theory*

This approach is based on the observation that an economic agent, consumer or producer, generally prefers to consume a good or service immediately rather than in the future. Symmetrically, the economic agent prefers to receive 1 euro now rather than later. This fact can be interpreted by associating the concept of discount rate with the "price of time". In practice, this effect can be evaluated by an estimate of the interest rate to be applied so that household consumption can remain indifferent to dates (consumption today, or at a later moment).

This approach projects an individual psychological effect at the level of the community. One can define a social discount rate representing the value society places on time. The rate of pure preference for the present δ intervenes in Ramsey's equation (1928) and provides an analytical expression of the discount rate τ:

$$\tau = \delta + \gamma g \qquad [2.10]$$

Terms γ and g, respectively, represent the elasticity of the marginal utility of consumption and the growth rate of individual income. The reader will find a recent demonstration of this formulation in Gollier and Hammit (2014).

In the economic debate, the existence of a rate of pure preference for the present is subject to controversy. A 1989 study by Benzion et al. (1989) based on the participation of 204 students refers to this. One of the study's conclusions calls into question the existence of a single, individual discount rate, independent from the context, the time periods or amounts involved. The interpretation of this study by Frederick et al. (2002) suggests it is possible for the pure time-preference rate to be zero after controlling certain exogenous variables. In addition, the transposition of individual psychological behavior to society as a whole is also called into question.

2.4.1.2. *Net marginal productivity of capital*

This approach can be demonstrated by the contest between consumption and investment. In this section, we reproduce the logic of the approach as detailed in French Planning Agency (Commissariat général du plan 1968). At the community level, an investment $\Delta I(t) > 0$ on date t can be seen as a levy on consumption $\Delta C(t) < 0$ on this same date. This investment is economically viable as long as it provides an increased consumption $\Delta C(t+1) > 0$ in the future. The discount rate is then the rate which expresses this equivalence:

$$\tau(t) = \frac{\Delta C(t+1)+\Delta C(t)}{-\Delta C(t)} \qquad [2.11]$$

For example, if a community is willing not to consume €1 on date t in exchange for a consumption of 1.1 € on date $t+1$, the social discount rate is 10%.

We call $\Delta I(t)$ the potential investment whose economic relevance we wish to assess. Let us consider an aggregate production function whose form is $Y = F(K, L)$, where K and L represent capital and labor. It is then possible to determine the increase in the aggregate production at instant $t+1$ generated by this investment:

$$\Delta Y(t+1) = \frac{\partial Y}{\partial K}(t) \cdot \Delta I(t) \qquad [2.12]$$

The term $\frac{\partial Y}{\partial K}(t)$ represents the marginal productivity of capital. However, the investment $\Delta I(t)$ can be achieved at the cost of an equivalent decrease in consumption. So we have:

$$\Delta \mathrm{I}(\mathrm{t}) = -\Delta C(t) \qquad [2.13]$$

Then, the increase in the aggregate production generated by the investment is expressed as follows:

$$\Delta Y(t+1) = -\frac{\partial Y}{\partial K}(t) \cdot \Delta C(t) \qquad [2.14]$$

This increase in the aggregate production can be broken down into an increase in consumption $\Delta C(t+1)$ and a reduction in the investment of an amount equal to $\Delta I(t)$, corrected for its depreciation at rate δ over the period. So we have:

$$\Delta Y(t+1) = \Delta C(t+1) - (1-\delta)\Delta I(t) \qquad [2.15]$$

By reinjecting this expression into the one above, we get the following equality:

$$\tau(t) = \frac{\partial Y}{\partial K}(t) - \delta \qquad [2.16]$$

The public discount rate is therefore equal to the difference between the marginal productivity of capital and the depreciation rate, by definition, the net marginal productivity of capital.

2.4.1.3. *Market interest rate*

The approach presented in this section is based on Gollier and Hammit's (2014) work. In a simplified way, the starting point for this approach is based on a firm seeking to evaluate a potential investment on a given date. If the remuneration exceeds the investment cost, then the project is considered economically profitable. This concept can be simply grasped by reasoning about a non-risky project. The firm offers lenders a remuneration at an interest rate close to that of the market. To generate a profit, the project's remuneration must exceed this rate. Firms are encouraged to use the risk-free interest rate as discount rate to evaluate non-risky projects. In practice, a risk allowance is added. Above all, private companies take into account the access conditions to capital (and the trade-off between equity capital and loans), which modifies their exposure to risk and as a consequence, the discount rate.

In the case of a public discount rate, the analogy between the discount rate and the interest rate has a number of limitations. Financial assets have due dates which rarely exceed 30 years. For such a timeframe, the risk is potentially very high (risk of default, inflation, etc.). Moreover, the uncertainty of long-term public calculation must incorporate the growth of the agents' incomes.

This shows why the discount rate will become an elusive concept to handle in a few decades. It should be noted that energy investment projects, particularly in the nuclear field, have relatively high timeframes.

2.4.2. *Values and effects of the discount rate*

2.4.2.1. *The rates used for the nuclear sector*

There is a large literature on considerations related to the discount rate value to be applied when evaluating public investment projects. Thus, the recommendations from Charpin et al.'s (2000) report lead to a variable discount rate, depending on the chosen time horizon:

– 0–30 years: 6%;

– over 30 years: 3%.

In Lebegue et al. (2005), the discount rate for public investments is 4%, and then decreases over time after 30 years with a minimum value set at 2% in the context of very long-term investments.

In its report on the nuclear sector published in 2012 (Cour des comptes 2012), the Court of Auditors called the attention on "the lack of consensus on the discount rate to be used". The report details the calculation method used by the nuclear industry and highlights the fact that the discount rate used is 5%.

In 2013, the Quinet commission (Quinet and Baumstark 2013) proposed "placing the evaluation of investments in the list of long-term problems". The commission endorsed the idea that the current context is characterized by strong uncertainty and "does not allow us to extrapolate the future by assuming that development will take place in a homothetic way". The scenario approach is recommended while specifying that economic criteria are not sufficient to assess long-term investment projects. This context of uncertainty is in line with a reduction in the public discount rate. The latter must be set at 2.5% for the next few years and should gradually decrease to 1.5% in the future beyond 2070. This rate is increased by a risk allowance of 2% weighted by a coefficient calculated for each project and which depends on "its profitability sensitivity to economic growth". This allowance increases from 2% to 3% beyond 2070.

In practice, given the absence of evidence of a single rate, it is possible to carry out a study on the sensitivity of economic calculation to the discount rate. For example, the OECD/NEA studies presented in OECD/NEA (2020a) offer results with a 5% fixed rate, and then a 10% fixed rate.

The major challenge of choosing a rate is the distinction between the public actor (the State) and the other actors (industrialists, bankers in particular). As already observed, these actors do not have the same role as the State and the risks they have to face are generally greater (if only because their "size" is smaller and the risk of bankruptcy can be much greater for large-scale projects).

In general, private rates (around 5–10%) are higher than public rates (around 3–5%). While a project can be interesting for the community (positive social NPV), this may not be the case for the private sector (for the investor). States must therefore correct this situation, in particular by offering guarantees to the investor for coping with the project's risk (this subject is illustrated in Chapter 3).

The case study discussed in section 2.5 will be based on several discount rates presented in the literature. Two rates (4% and 7%) will be used to represent public and private choices.

2.4.2.2. *Discount factors*

The discount factor is defined as the function weighing a cash flow $c(t)$ executed on date t. In practice, there are continuous and discrete approaches. A cash flow vector can thus be updated via the continuous discount rate R:

$$F_c = \int_{t_1}^{t_2} c_c(t)\, e^{-Rt} dt \qquad [2.17]$$

In this expression, F_c represents the present value of the cash flow and $c_c(t)$ is the continuous cash flow per time unit. The cash flow vector can also be expressed via a discrete approach inferred from a discreet discount rate:

$$F_d = \sum_{t=t_1}^{t=t_2} \frac{c_d(t)}{(1+r)^t} \qquad [2.18]$$

In this case, F_d represents the present value of the cash flow and $c_d(t)$ is the discrete cash flow per time unit. At the boundary between the continuous and discrete approach, considering an infinitely small time unit, we can impose $F_c = F_d$ and equalize cash flows $c_c(t)dt = c_d(t)$. We then obtain an equivalence between rates in the following manner:

$$R = \ln(1 + r) \qquad [2.19]$$

2.4.2.3. *Evolution of discount factors*

With the equivalence between the continuous and discrete discount rate defined in the previous section, it is now possible to outline these functions. Figure 2.3 represents this function for some fixed rates.

Over the period considered, rather representative of a "nuclear temporality", the discount factor has a strongly decreasing value between 0 and 1. Thus, a cash flow in 60 years' time discounted at t = 0 is weighed by a discount factor of 0.55 at 1%, and 0.017 at 7%. It follows that the economic values of nuclear programs critically depend on the first 20–30 years.

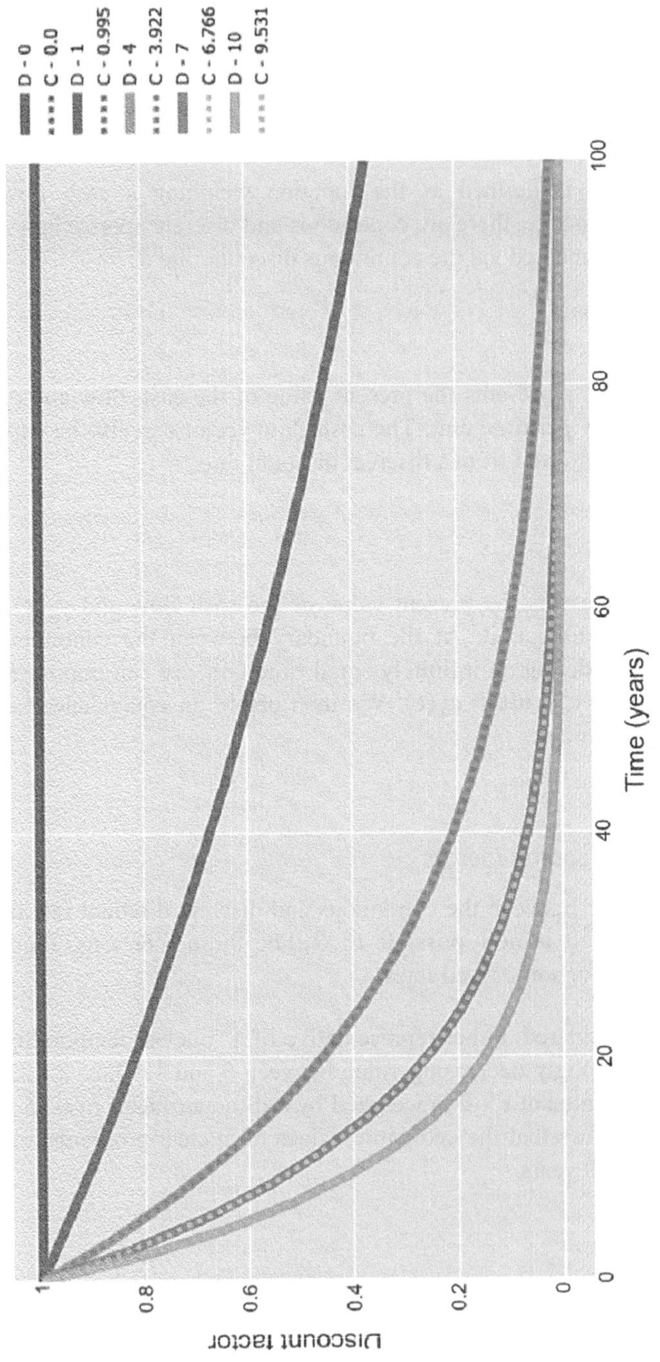

Figure 2.3. *Evolution of the discount factor as a function of time over a time period of 100 years for five discrete rates (D) and their continuous equivalents (C): 0.1%, 4%, 7% and 10%. For a color version of this figure, see www.iste.co.uk/percebois/nuclear1.zip*

2.5. Case study: a new nuclear reactor

In this section, we will introduce the reference scenario for which the average discounted cost will be calculated. From this reference calculation, a sensitivity analysis will highlight the main parameters of interest.

2.5.1. Reference scenario

The reference scenario is based on data whose values are justified in detail in Chapter 3[11].

Here, the calculation is based on a scenario for a reactor whose technical characteristics are clearly defined. The time reference will be the reactor's year of commissioning.

2.5.1.1. *Technical characteristics of the reactor*

The type of reactor chosen will be an EPR with a nominal electrical power of 1,650 MW and a load factor by 90%. The latter depends not only on the number and duration of partial shutdowns, but also on network constraints. The main determinants of the load factor are then:

– refueling;

– maintenance operation;

– ten-year inspections;

– load monitoring.

The reactor's lifetime has been set at 60 years, with upgrade work between 40 and 50 years.

2.5.1.2. *Construction phase*

The reference construction schedule is set at 6 years. The reactor's overnight cost is €4,500/kW and can be broken down into the following components, quantified in order of magnitude according to D'Haeseleer (2013):

– engineering, supply and construction (71%);

11 The source codes for the LCOE calculations implemented in, in this section are freely available at the following address: https://github.com/thiollie/lcoe.

– owner costs (18%);

– contingencies (11%).

The expenditure time profile during the construction follows a slightly left-skewed normal law. Although each construction is unique, this distribution is intended to be representative of a "realistic" industrial schedule.

In addition to the overnight costs CCO, there are the interests during construction) IIC (IDC, also called interim interests, which need to be considered when calculating the total investment cost CTI. The interests during construction are sometimes referred to as "financial costs". They represent the interest paid during the construction phase of the plant. Not only do they include the debt component of the investment, but also to the investors' remuneration.

The total cost of construction is calculated as the sum of the overnight cost and the interim interest. The total construction cost can be calculated as follows (European Commission 2008):

$$CTI = CCO \sum_{t=1}^{t=CT} \frac{W_t}{(1+r)^{(t-1-CT)}} \qquad [2.20]$$

In this equation, CT represents the construction schedule, W_t is the overnight cost fraction spent per year t and r represents the discount rate.

Interim interest can also be calculated in a more intuitive way, by summing at each stage the interest calculated in relation to the cumulative loan from the previous years. This method is fully described in D'Haeseleer (2013, p. 21).

Figure 2.4 represents the expenditure history of the reference project according to the above-mentioned hypotheses, showing the construction costs, as well as the interim interest for two different rates. By convention, the IIC have been assigned to the beginning of each time period, while production starts in the middle of year 0.

In this example, the total construction cost is €7.425 billion with interim interest amounting to €1.0 billion for a 4% rate, and €1.9 billion for a 7% rate.

2.5.1.3. *Production stage*

The electricity production phase is effective from time 0 until the end of the year 59. The denominator in the LCOE formula shows an energy production term weighed by the discount factor. This formulation may imply that the economic calculation discounts physical units of energy, which is obviously not correct. This point has been addressed in OECD/NEA (2010).

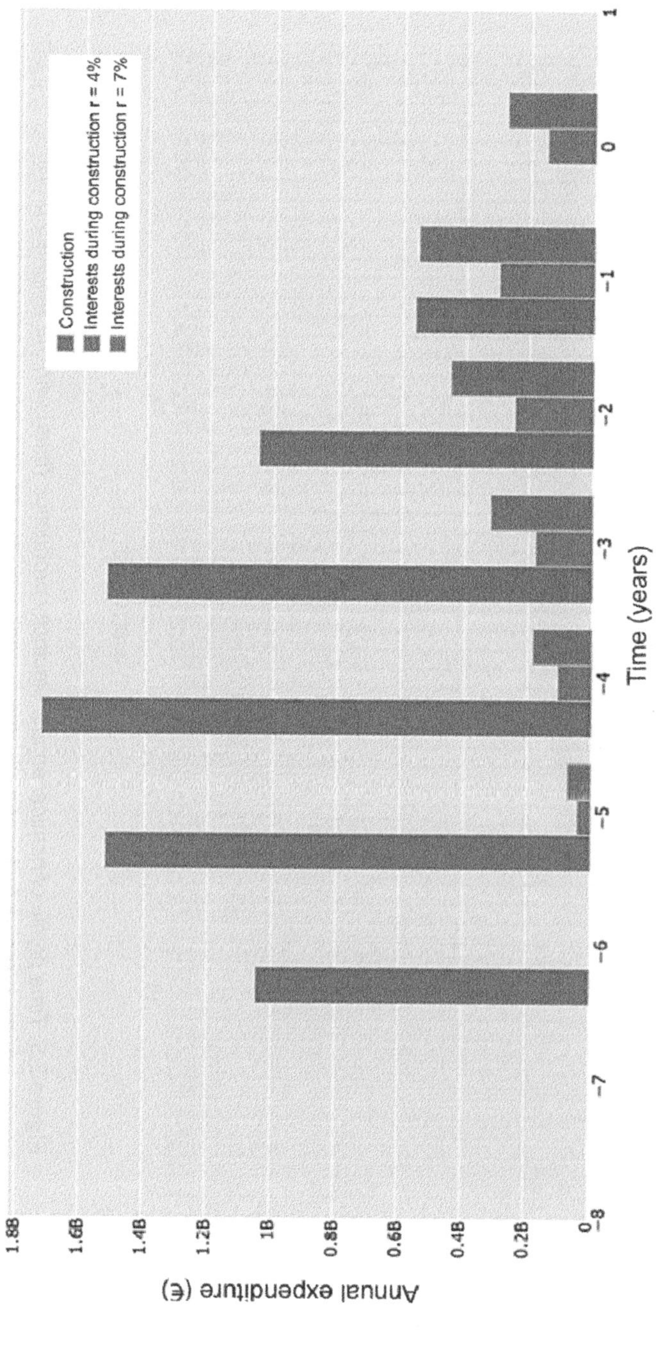

Figure 2.4. *Profile of construction costs and interim interest calculated for a 7% and 4% discount rate. For a color version of this figure, see www.iste.co.uk/percebois/nuclear1.zip*

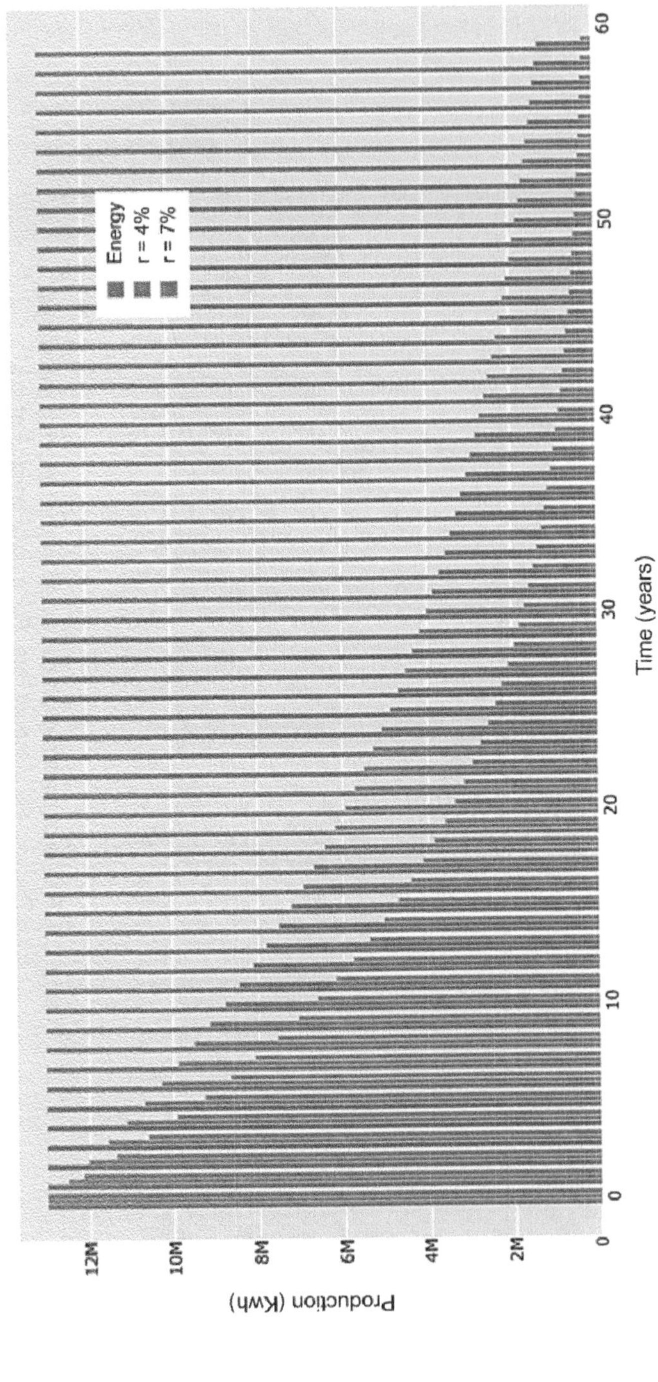

Figure 2.5. *Time profile of nominal annual electricity production (blue) with the weighted profile of the discount factor at a 4% rate (in red) and 7% rate (in green). For a color version of this figure, see www.iste.co.uk/percebois/nuclear1.zip*

Energy is actually integrated into the calculation through the value of its sale at a constant price, which is materialized by a discounted cash flow. The resulting price from the sum is the product of the energy by the discount factor.

Figure 2.5 represents the nominal annual quantities of electrical energy produced as well as those weighted by the discount factor. The energy profile clearly shows the influence of the discount rate on the LCOE's denominator.

2.5.1.4. *Input economic data*

At this stage, it is possible to represent the cash flow over the entire project's life. These disbursements are calculated on the basis of the data presented in Table 2.1.

These economic data are based on the chapter relating to the costs of nuclear energy (Chapter 3).

	Investment (€/kW)	Fuel (€/MWh)	O&M (€/MWh)	R&D (€/MWh)	Refurbishment (€/MW)	Decommissioning (% of construction overnight cost)
Amount in $€_{2020}$	4,500	8	13	2	600	15

Table 2.1. *Summary of the costs used for the reference calculation*

The construction item represents the construction expenses and the calculated interim interest via the discount of expenses for negative times. This representation of the interim interest is only visual and helps to understand the influence of the discount rate on the total construction expenditure. The fuel costs are set at €8/MWh. They take into account the management costs as well as the provisions dedicated to the management of spent fuel. Operation and maintenance costs are essentially variable costs set at €13/MWh. We have also included R&D costs at €2/MWh. Disbursements relating to refurbishment costs appear after 40–50 years for a total amount of €600/kW. Finally, disbursements related to decommissioning are represented after 60–70 years with a total cost representing 15% of the construction overnight cost.

2.5.2. *Results*

This section details the results of the LCOE calculation and introduces a sensitivity study to the economic input data.

2.5.2.1. *Disbursement profile*

The disbursement profile is shown in Figure 2.6. All the positions listed in Table 2.1 are represented. Three discount rates were used to construct this figure. The zero rate represents nominal cash flows. Rates of 4% and 7% were also used, providing good understanding of the discount effect on long-term flows.

Note that the expenses incurred after 30 years are "crushed" and will only take part slightly in the calculation of the LCOE. This fact also helps to understand why decommissioning is an item whose contribution to the LCOE remains low, regardless of the cost hypotheses chosen.

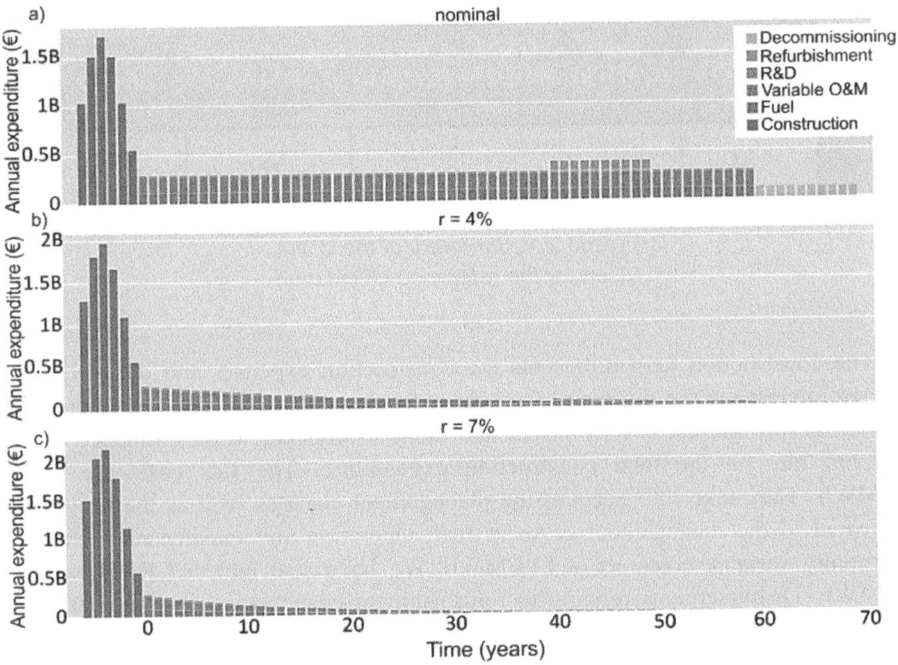

Figure 2.6. *Disbursement for a new EPR-type nuclear reference project. (a) a zero discount rate; (b) a 4% rate; (c) a 7% rate. For a color version of this figure, see www.iste.co.uk/percebois/nuclear1.zip*

2.5.2.2. *LCOE calculation and its main components*

The discounted cost of electricity is calculated using formula 2.4. In practice, this means summing the discounted cash flows and dividing them by the sum of the energy production multiplied by the discount factor. We then obtain a €52/MWh LCOE for a 4% rate and €76.6/MWh for a 7% rate. The breakdown of these LCOEs is detailed in Figures 2.7 and 2.8.

First of all, note that the cost of capital is the largely predominant component for the two discount rates used. The reactor's overnight cost is €7.4 billion and the CAPEX costs are, respectively, €8.5 billion and €9.3 billion for the 4% and 7% discount rates.

The increase in the discount rate has several effects. First, it increases the cost of capital through an increase in interim interests. Second, this reduces discounted operating costs and decommissioning costs. Finally, this decreases the denominator of equation [2.4] via the discount factor.

Ultimately, these cumulative effects imply an increase in LCOE with the discount rate. It follows that the discounted costs are higher for an electricity supplier, compared to those taken into account by the State.

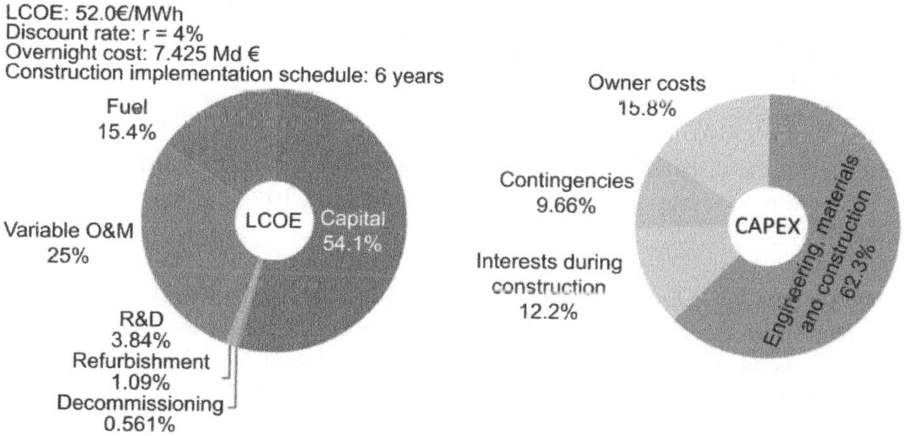

a) b)

Figure 2.7. *Breakdown of LCOE for a 4% discount rate (a). Breakdown of capital expenditure (b). For a color version of this figure, see www.iste.co.uk/percebois/nuclear1.zip*

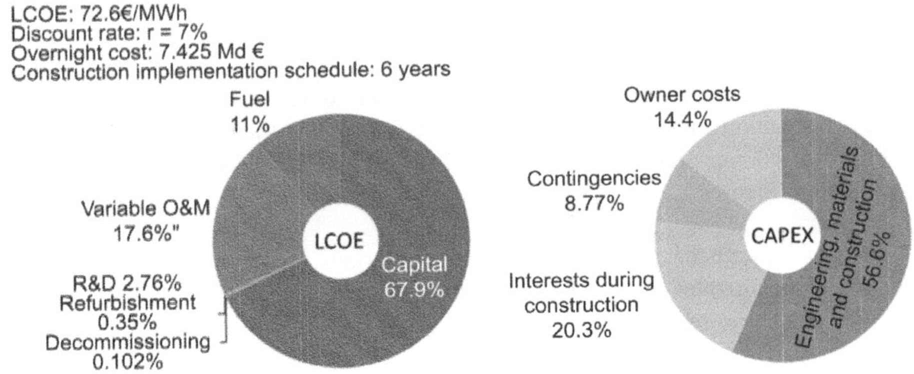

a) b)

Figure 2.8. *Breakdown of LCOE for a 7 % discount rate (a). Breakdown of capital expenditure (b). For a color version of this figure, see www.iste.co.uk/percebois/nuclear1.zip*

2.5.2.3. *Sensitivity analysis*

This section intends to analyze the sensitivity of the reference LCOE calculated in the previous section to input economic data. First, we propose a discount rate variation between 0 and 10%. The dependence of the LCOE on the discount rate is shown in Figure 2.9.

We come across the result from the previous section. The LCOE increases sharply with the discount rate in a nonlinear way. The latter varies from €35/MWh for a zero discount rate up to €100/MWh for a constant 10% discount rate over the project's life.

We now propose to analyze the 10-year reduction in the operating period, which also results in the removal of the upgrade phase at $t = 40$ years. The lifespan therefore decreases from 60 to 50 years. After calculation, we obtain €53.1/MWh (compared to €52/MWh in the reference scenario) and €73.3/MWh (compared to €72.6/MWh in the reference scenario) LCOEs for the 4% and 7% discount rates. The cumulative effects of these two factors are relatively low for the two discount rates chosen.

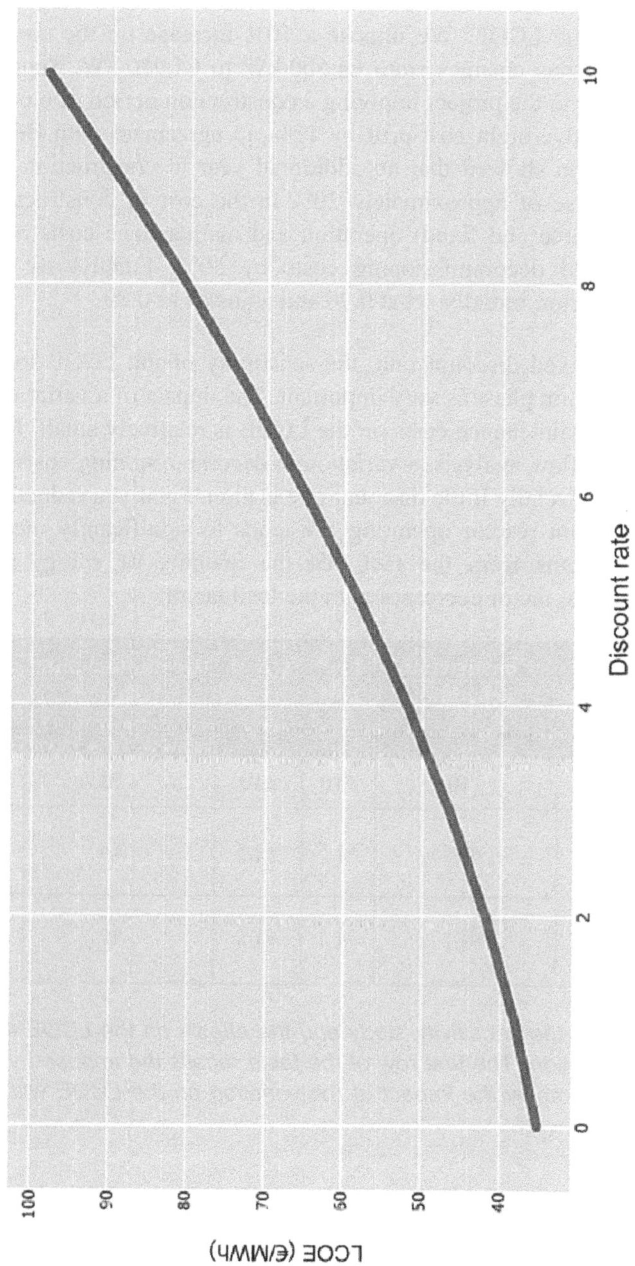

Figure 2.9. *Representation of the LCOE depending on the discount rate assumed to be constant over the entire project's life. For a color version of this figure, see www.iste.co.uk/percebois/nuclear1.zip*

Finally, we suggest modifying the input parameters separately, in order to isolate the individual contribution of each term. Table 2.2 summarizes the changes made as well as the effect on the LCOE. We impose a 10% increase on the construction overnight cost, which thus changes from €4,500/kW to €4,950/kW. Separately, a 10% increase is applied to the project implying a construction period of 6.6 years, as well as a construction overnight cost drift by 12%, in agreement with Berthélémy and Escobar (2015) who showed that an additional year in construction schedule could induce an increase of approximately 20% in the cost of construction. In a separate manner, (variable and fixed) operation and maintenance costs have been increased by 10%, and decommissioning costs by 50%. Finally, we tested a reduction in the load factor, initially set at 0.90 and adjusted to 0.80.

We find that at a fixed discount rate, the sensitivity of the LCOE to the data relating to the construction phase is very important. The impact of a variation in fuel cost or operation and maintenance costs on the LCOE is relatively small. By means of the discounted cash flow analysis, a variation in decommissioning costs has little effect on the LCOE, especially if the discount rate is high. Finally, a reduction in the load factor, at a constant reactor operating life tends to significantly increase the LCOE. This effect stems from the fact that the quantity of energy produced weighted by the discount factor decreases with the load factor.

	Construction cost (%)	Construction schedule (%)	Fuel cost (%)	O&M cost (%)	Decommissioning cost (%)	Load factor (%)
Variation	+10	+10	+10	+10	+50	–10
LCOE (r=4%)	+5.6	+6.9	+1.5	+2.5	+0.4	+6.9
LCOE (r=7 %)	+6.7	+8.7	+1.1	+1.8	~0	+8.5

Table 2.2. *Summary of the sensitivity study and the effects on the LCOE in relation to the reference calculation. The first row of the table recalls the imposed variations. The following two rows show the impact of the variation on the LCOE with 4% and 7% discount rates*

2.6. Conclusion

As it has been shown, the field of economic analysis of nuclear energy has expanded. In the 1980s, the problems studied were almost exclusively related to the choice of electricity production means. The focus was on the calculation of discounted costs for future projects, paying particular attention to the fuel component. In France, the work by Bertel and Naudet (2004) was focused on these same questions, but added new problems of several kinds: externalities (including climate), the energy policy, with a passage from an approach choosing a production means to an entire fleet, the macroeconomic impact. The underlying problem was mainly centered on future investment choices, made either by the government or by a nationalized producer applying government directives. The reactors concerned were pressurized water machines relying on Framatome technology, and then Areva. In all of the cases, the French-speaking research works addressing these questions were largely focused on French questions.

The journey toward the "generalization" of economic approaches along the three "axes" that were identified at the beginning of this chapter had been underway for decades. Despite everything, the international economic literature on nuclear energy is relatively limited, compared to that on renewable energies[12]. But it is nonetheless sufficiently extensive for pretending to summarize it in only one work, in particular regarding the question of reactor costs in the broad sense (types of reactors and cost categories).

Still, the production cost of nuclear-generated electricity is the "core" (in the sense of Figure 2.1) of this entire economy. As the subject of investment in new capacities is increasingly present in the agenda of various countries, particular attention has been paid in this chapter to the LCOE and its foundations.

This tool, as well as many others which have been briefly described at the beginning of this chapter, exist only because questions arise, to which economists can provide some answers. We have already mentioned some of them and will summarize the main ones below:

– How to evaluate past choices (historical nuclear programs)?

– Is nuclear power competitive in the daily operation of the current fleet?

– How to price electricity taking into account the real historical cost of nuclear energy?

– Should we economically extend the operating life of existing reactors?

– Are the bases for the choices of tomorrow's electricity exhaustive? What are their possible hidden costs? How is nuclear energy positioned in relation to other production means?

---

12 Less than 100 papers per year versus several thousand for renewable energies.

– How to reduce future nuclear costs?

– What type of “low carbon” electricity production for the future? Is a new nuclear program in Europe (in France) economically justified?

Chapter 3 attempts to provide a summary of results, shedding light on these different questions.

2.7. References

Benzion, U., Rapoport, A., Yagil, J. (1989). Discount rates inferred from decisions: An experimental study. *Management Science*, 35(3), 270–284.

Bertel, E. and Naudet, G. (2004). *Économie de l’énergie nucléaire*. EDP Sciences, Les Ulis.

Berthélémy, M. and Escobar, L. (2015). Clear reactors’ construction costs: The role of lead-time, standardization and technological progress. *Energy Policy*, 82, 118–130.

Brook, B.W. and Bradshaw, C.J.A. (2014). Key role for nuclear energy in global biodiversity conservation. *Conservation Biology*, 29(3), 702–712.

Charpin, J.-M., Dessus, B., Pellat, R. (2000). Étude économique prospective de la filière électrique nucléaire. Report for the Premier ministre.

Commissariat général du Plan (1968a). Taux d’actualisation et rationalité économique. Première partie. Report, French General Planning Commission.

Commissariat général du Plan (1968b). Taux d’actualisation et rationalité économique. Deuxième partie. Report, French General Planning Commission.

Cour des comptes (2012). Les coûts de la filière électronucléaire. Report, Cour des Comptes, Paris.

D’Haeseleer, W.D. (2013). Synthesis on the economics of nuclear energy. Study for the European Commission, DG Energy.

Devezeaux, J.-G. and Berthélémy, M. (2020). Coûts du nouveau nucléaire et éclairage sur l’économie du cycle. *Annales des Mines, Responsabilité et environnement, Le nucléaire civil, enjeux et débats*, 97, 35–47.

European Commission (2008). Energy sources, production costs and performance of technologies for power generation, heating and transport. European Commission, Brussels.

Fery, P. (2010). Risque et calcul socio-économique. Report, Centre d’analyse stratégique, Paris.

Fisher, I. (1930). *The Theory of Interest*. Macmillan, New York.

Frederick, S., Loewenstein, G., O'Donoghue, T. (2002). Time discounting and time preference: A critical review. *Journal of Economic Literature*, 40(2), 351–401.

Gollier, C. (2011). Le calcul du risque dans les investissements publics. Report, Centre d'analyse stratégique, Paris.

Gollier, C. and Devezeaux de Lavergne, J.-G. (2001). Analyse quantitative de la réversibilité du stockage des déchets nucléaires : valorisation des déchets. *Économie & prévision*, 149, 1–13.

Gollier, C. and Devezeaux de Lavergne, J.-G. (2002). Valeur de la réversibilité : analyse quantitative de la réversibilité du stockage des déchets nucléaires. *Économie et Prévision*.

Gollier, C. and Hammit, J.K. (2014). The long run discount rate controversy. *Annual Review of Resource Economics*, 6, 273–295.

Harvey, C. (1989). Prescriptive models of psychological effects on risk attitudes. *Annals of Operations Research*, 19, 143–170.

Hirth, L., Ueckerdt, F., Edenhofer, O. (2015). Integration costs revisited – An economic framework for wind and solar variability. *Renewable Energy*, 74, 925–939.

IAEA (2019). Application of multi-criteria decision analysis methods to comparative evaluation of nuclear energy system options: Final report of the INPRO Collaborative Project KIND.

Koopmans, T. (1960). Stationary ordinal utility and impatience. *Econometrica*, 28, 287–309.

Lebègue, D., Baumstark, L., Hirtzman, P. (2005). Révision du taux d'actualisation des investissements publics. Report, Commissariat général du Plan.

OECD/NEA (2010). Projected costs of generating electricity. Available at: https://www.oecd-nea.org/upload/docs/application/pdf/2019-12/6819-projected-costs.pdf.

OECD/NEA (2012). Nuclear energy and renewables: System effects in low-carbon electricity systems. Available at: https://www.oecd-nea.org/upload/docs/application/pdf/2019-12/system-effects-exec-sum.pdf.

OECD/NEA (2015). Projected costs of generating electricity. Available at: https://iea.blob.core.windows.net/assets/c9bae6ac-0f4c-4a4b-8b46-f7d4cca4d53b/ElecCost2015.pdf.

OECD/NEA (2019). The costs of decarbonisation: System costs with high shares of nuclear and renewables. Available at: https://www.oecd.org/publications/the-costs-of-decarbonisation-9789264312180-en.htm.

OECD/NEA (2020a). Projected costs of generating electricity. Available at: https://www.iea.org/reports/projected-costs-of-generating-electricity-2020.

OECD/NEA (2020b). Unlocking reductions in the construction cost of nuclear. A practical guide for stakeholders. Available at: https://www.oecd-nea.org/upload/docs/application/pdf/2020-07/7530-reducing-cost-nuclear-construction.pdf.

Portney, P.R. and Weyant, J.P. (eds) (1999). Introduction. In *Discounting and Intergenerational Equity*. Routledge, New York.

Quinet, É. and Baumstark, L. (2013). L'évaluation socioéconomique des investissements publics. Report, Commissariat général à la Stratégie et à la Prospective.

Ramsey, F. (1928). A mathematical theory of saving. *The Economic Journal*, 38(152), 543–559.

de Roo, G. and Parsons, J.E. (2009). Nuclear fuel recycling, the value of the separated transuranics and the levelized cost of electricity. CEEPR Report 09-008.

de Roo, G. and Parsons, J.E. (2011). A methodology for calculating the levelized cost of electricity in nuclear energy systems with fuel recycling. *Energy Economics*, 33(5), 826–839.

Rothwell, G. (2015). *The Economics of Future Nuclear Energy*. Routledge Press, Chicago.

Samadi, S. (2016). A review of factors influencing the cost development of electricity generation technologies. *Energies*, 9(11), 970.

Samadi, S. (2017). The social costs of electricity generation – Categorising different types of costs and evaluating their respective relevance. *Energies*, 10, 356.

Samuelson, P. (1937). A note on measurement of utility. *Review of Economic Studies*, 4, 155–161.

Schubert, K. (2006). Éléments sur l'actualisation et l'environnement. *Recherches économiques de Louvain*, 72, 157–175.

Taverdet-Popiolek, N. and Shoai Tehrani, B. (2015). Évaluation économique de la R&D relative aux réacteurs rapides de génération IV : le cas de la France. *Revue de l'énergie*, 627.

3

The Production Costs of Nuclear Electricity

Jean-Guy DEVEZEAUX DE LAVERGNE
Formerly I-tésé (CEA), Gif-sur-Yvette, France

3.1. Introduction

The notions of cost were introduced in Chapter 2. We now have most of the tools needed to describe and evaluate the costs of historic nuclear power as well as those of future nuclear power.

We will explore historical costs (Generation II) in detail, something which constitutes a clear step ahead compared to previous works: in recent times, the economic literature has extensively developed around this topic, particularly in France. We will briefly review this work for methodological purposes. Then we will examine the various levels of the concept of "historical cost" around the world and in France.

We will then review the costs of the nuclear revival (Generation III), which is a much debated topic in many countries.

Finally, we will evaluate system costs, whose recognition has revolutionized the economics of electrical systems over the past 10 years. To conclude, we will illustrate the concept of total cost (or full cost) which takes into account the whole of quantifiable costs apart from the discounted cost from a public perspective. This approach is essential for decision-making.

Nuclear Economy 1,
coordinated by Jacques PERCEBOIS and Nicolas THIOLLIÈRE.

3.2. Nuclear costs of existing reactors (Generation II)

These costs relate to investments in reactors (and nuclear cycle plants), which are currently in service. Many studies document them at a global scale (see below). A specificity of nuclear energy is the significant weight of costs a priori from past electricity generation: reactor decommissioning, retrieval and packaging of "historic" waste, and long-term management of waste still devoid of a specially dedicated facility. These costs, which are closely regulated by law, are evaluated, monitored and made public according to specific procedures. They remain shrouded in a level of uncertainty typically ranging between 10 and 30%. As shown in Figure 2.8, their weight is nonetheless weak, in particular as soon as the usual discount rates are applied.

3.2.1. *Cost assessment methods for historical reactors*

Making the historical costs of reactors economically consistent involves three types of difficulty: differences in nomenclature and scope, economic consideration of variations in monetary parity and awareness about the passing of time (inflation, possible cost of capital and discount). A comparative analysis requires an approach by country (specific costs, organization, legislation, etc.), which aims to account for the differences and developments. The methods for comparing reactors are inspired in accounting analysis, whereas the corrections to be applied mainly arise from the engineering field. Cost deflators are also a crucial issue, as are exchange rates. Finally, standard economic approaches (such as econometrics) can make data more significant in time and space, for example, by pondering the influence of groups of countries or types of reactors.

3.2.1.1. *Cost calculation methods*

This subject is "technically" complex. As much as it is almost universal to evaluate the cost of an investment (with more or less detail, as indicated above), determining a production cost can result in multiple approaches. Furthermore, Vattenfall, TVA or EDF have their own analytical cost calculations, which are not disclosed to the general public. But these calculations, which aim to determine the profitability of a company's internal activities or to anticipate and "steer" the annual result, are of little interest for economic analysis, as they are far too specific and mainly focus on the short-term.

For the most part, the analysis of past costs is analytical in nature, but this work also falls largely within a public field. The difficulties associated with these types of evaluation essentially concern the proper definition (and understanding) of the method used. The major point concerns taking into account (or not) the costs of the capital in place, or even the introduction of a rule of return on capital aimed at obtaining a cost that is consistent with a predefined level of economic performance[1]. Taking time into account can also become complex by associating capital already deemed to have been more or less amortized with future investments. In the latter case (essentially the extension of operation), costs can be determined for periods ranging from one accounting year to around 10–15 years.

3.2.1.1.1. Short-term variable cost

This concept of cost, the lowest in €/MWh of the list below, is directly proportional to the energy produced. This is the short-term cost of an additional supply of one MWh, which depends on the moment when it is requested. When nuclear power is marginal (meaning that the last unit required is nuclear, or the whole available nuclear fleet is not fully used), the marginal cost of production is equal to this short-term variable cost that fixes the market price at this moment.

It is essentially used for managing the fleet in the short-term (summoning the reactors or not). It is considered as cash, because it corresponds to disbursements made at the time of the production (fuel[2], intermediate consumption, but also staff needs). The existing nuclear power has some of the lowest short-term marginal costs (most nuclear costs are fixed). Such a level ensures that nuclear energy is generally summoned. But the significant increase in the power of renewable energies (wind and solar, but sometimes also hydraulic in Northern Europe countries) can lead to this equipment fixing the market price at levels close to zero (or even negative). In such conditions, it may happen that nuclear energy is not required at all (see Chapter 2, Volume 2).

This short-term cost is "perceived" at each power station terminal and is defined by the technical reality of the equipment. There is also a long-term marginal cost (or development cost), which corresponds to the same definition, but includes the investment cost, assuming that the fleet's structure and level can be adapted to future demand. This concept is the basis for the Ramsay–Boiteux pricing theory (Percebois 1989). It is "interpreted" from the demand's perspective: the focus is on the cost

1 This rule is only loosely related to the fact that productive capital has been more or less amortized in the company's accounts. Depreciation is used for calculating the result for the company, and the taxing scheme (corporate tax). For example, capital can be quickly amortized for tax purposes (with a degressive scheme, for example), whereas the *cash flows* from investment have remained modest and economic balance is far from being achieved. Social accounting often has little to do with economics.

2 See Bertel and Naudet (2004) for whom a portion of the nuclear fuel cost is fixed (not proportional to short-term production).

caused by the incremental demand, once the fleet has been optimized. This is not an issue related to a specific production means, and this notion is not exclusively applicable to nuclear power. On the other hand, the cost and performances (flexibility, etc.) of future nuclear power influence the calculation of all the long-term marginal costs, as well as all the other production means and balancing methods.

3.2.1.1.2. Full production cost

The calculation of the "full" production costs of historic nuclear power is key to answering many questions. It is a family of concepts quite close to one another, which differ in the approach to capital, either with an accounting (depreciation) or an economic logic (based on remuneration). Three of these concepts will be presented in the next section. The first one is purely accounting related. The last two have recently been used in France in a context in which the electricity market is becoming increasingly open, with a view to enabling new entrants to compete with the traditional operator (EDF), which has historically profited from low production costs acting as a barrier to the entrance. This situation is the result of a relative advantage from past risk-taking of nuclear investments that have now become profitable, in which they have been largely amortized. The idea is to enable new entrants to obtain supplies from EDF at purchase prices reflecting production costs. This approach was the basis for the design, and the subsequent implementation of the so-called *ARENH* tariff (regulated access to historical nuclear electricity), described in detail in Chapter 2, Volume 2.

Full accounting production cost

This concept is derived from an analytical accounting logic, following a "profit and loss" account structure, but breaking down costs per component, as described above. In its simplest version, it comes down to adding the various current components and then taking into account the capital via tax rules for social accounting amortization. There is no question of remuneration for this capital. A different full cost is obtained every year. As this approach makes little economic sense, a more elaborate version has been defined by the French Court of Auditors (Cour des comptes 2012). In addition to breaking down operating and fuel costs by component, this approach includes capital via its amortization in accordance with accounting rules, while acknowledging the effect of inflation. Accounting amortizations are made on the basis of the value of investments on the date of completion. The Court's concept corrects endowments by assuming that investments are brought to present value. In addition to the return on capital not having been amortized in accounting terms, the calculation explicitly includes an "over-amortization" component related to the impact of inflation (extra-accounting) on the fleet's value. This concept can be considered as a basic tool for assessing the performance of the existing fleet, while trying to remain consistent with the social accounts of electricity suppliers.

Champsaur full cost

The question is different if one wishes to determine the cost price of the production sold, taking into account the return on capital economically (and no longer fiscally). In that case, we no longer refer to amortizations, or in any case, not only to amortizations, but to a reputedly "fair" return on capital. As it is not observable, it is necessary to define a rule to quantify it. It is possible to set a rate of return by regulation, and to directly infer a production cost, intended to serve as a guide for determining regulated tariffs. This is the case for many administrations in the United States. In France, the second committee chaired by Paul Champsaur (2012) chose to calculate an equivalent of the debt resulting from EDF's "nuclear activity" during the period studied and ensure that EDF would be able to reimburse via its *cash flow*. In this way, we can define a "fair" return on capital which has not yet been amortized. This concept aims to define a "fair price" for EDF's nuclear production costs during the period 2010–2025, that is to say, by ensuring that EDF will be able to meet the expenses of a major refit (operating life extension) over the period. This fair price is intended to serve as a reference for setting a tariff (that of the ARENH). Thus, the unit costs used include not only past investments (the fleet), operating costs, but also future investment costs to ensure the fleet's operation over time. This cost is assessed over a long period, but "fixed" to a specific year. It somehow presents a hybrid nature: while displaying an annual accounting perspective, it also offers a prospective vision. In a pragmatic and detailed way, the Champsaur commission verified the consistency of the results obtained with the regulatory practices of the time (the TarTAM, "ancestor" of the ARENH).

Current economic cost (Court of Auditors)

The Court of Auditors (2012, 2014) then sought to supplement the work of the Champsaur II commission regarding the question of the fair return on invested capital, but this time calculating it with several methods. We previously mentioned its definition of the Full Accounting Production Cost. But the Court also formalized the concept of economic current cost, which is more suitable for comparing the full cost of different energies. This concept makes no room for accounting depreciation. On the other hand, a remuneration of the capital is defined via a 7.8% fixed rate. This rate is expected to enable the operator to repay and remunerate investments over the entire operating life of the power plant, and to cover all of the past, present and future operating costs (including waste management-decommissioning). In spirit, the notion is close to the Champsaur cost and its purpose is similar as well: to define a reference tied as little as possible to past fluctuations. This reference can be compared to a tariff or sale price.

3.2.1.2. *Distinction between cash and non-cash costs*

Calculations of production costs essentially differ in relation to the methods for taking capital into account. In the social accounts, we access it via the gross operating surplus, ultimately, at the cash flow's source. In strict terms, this remuneration is not a cost component. To simplify, all the part of the turnover apart from the amounts disbursed for production and debt repayment (cash out) contributes to the cash flow. This in turn will make it possible to supply the financial assets on the balance sheet and to pay for new investments (for their self-financed portion). In approximate terms, the sale price level adjusted to strictly cover the amounts disbursed is the "dead point", beyond which the return on capital becomes positive.

This shows the advantage of making a difference (within the production cost) between the portion prompting an actual disbursement (here called *cash*) and the rest of the cost (non-cash), which is essentially a remuneration.

As already mentioned, there is a concept of short-term variable cost (typically per hour), which is essentially a cash cost. The full cash cost that we propose here differs from it in the sense that it is part of an annual period (as in business accounting, and in particular, the one determining the cash flows). So, in short, there are many distinctions among the cost concepts:

– A cash portion, corresponding to disbursements for the period and basically concerns operating expenses (labor costs and intermediate consumption, including fuel, taxes, etc.). Note that other cash flows of the period, such as financial flows, generally relate to past or future productions and must be restated.

– A non-cash portion, taking into account flows which have been or will be observed in other periods, but relate to the economic activity described: these are expenditures for investments already made and in place (whose cost is measured via different methods) and future expenses relating to the production of the period (decommissioning and waste management).

This distinction between cash and non-cash can apply to concepts such as accounting costs or current economic costs. It simply aims to identify what corresponds to disbursements at each moment for the electricity supplier in order to guide management decisions or to assess the performance of existing reactors in terms of cash flow.

3.2.1.3. *Conclusion on the choices of concepts and methods*

As it has been shown, the possible choices are diverse. Operationally, the various calculations lead to results that differ by several tens of percent. First and foremost, the difference between these production costs does not result from different input data on expenditure. It is also not the result of squabbles between schools over the application of a method allegedly being "the" correct one, for calculating "the true cost". It lies more in the diversity of the questions asked and in the definition of the relevant production cost, as in the way in which calculations must be carried out.

If we solely focus on the "historical" cost – assuming that the capital must simply be amortized in the accounting sense without being remunerated, without asking difficult questions about the reactors' operating life, without worrying about changes in the cost of reactors for a fleet renewal or significant fluctuation risks in future costs, and as long as we do not anticipate trends in the operation and the maintenance costs of aging facilities, for example – we can content ourselves with an accounting cost to assess the production cost. This can also be amended to take into account, at least in part, some of these points. One can also make the decision to refer to a certain cost and to check it each year so as to "stick as closely as possible" to the costs actually observed, but with the risk of remarkable fluctuations in the inter-annual result.

If, on the other hand, one wishes to take into account the reimbursement – or rather the remuneration – of the initial investment according to a stable signal, then the economic approach seems preferable (because it explicitly considers the remuneration of this capital). An important aspect of the choice of remuneration level concerns its consequences in terms of the amount of the implicit "bonus" attributed to the traditional electricity supplier for industrial management during the past decades, as electricity prices have long been regulated. Another aspect is the electricity supplier's ability (or lack of) to self-finance the fleet's lifetime extension (which is an issue for almost all the fleets in OECD countries), not to mention the renewal of this fleet.

There are two main reasons why these approaches give different results. These are, on the one hand, the capital-intensive nature of the nuclear industry and, on the other hand, the superposition of economic cycles (fleet age, interest rates, price index of production factors, etc.) and heavier trends signaling volatility and an increase in the prices of the majority of the components of the electricity tariff, be it of nuclear origin or not. In this sense, methods that give more weight to the past will lead to weaker present evaluations at the risk of probable later catch-ups that can be significant.

Thus, the "true" production cost of historic nuclear power is definitely inaccessible. Not because it remains hidden, but because it does not exist. There are real costs, which are well documented, known and debatable, determined within a clear framework (not to say transparent), adapted to different questions and treated in different ways. To objectify these remarks, concrete results will be provided in the following sections.

3.2.2. *Existing fleet costs*

3.2.2.1. *Construction Costs*

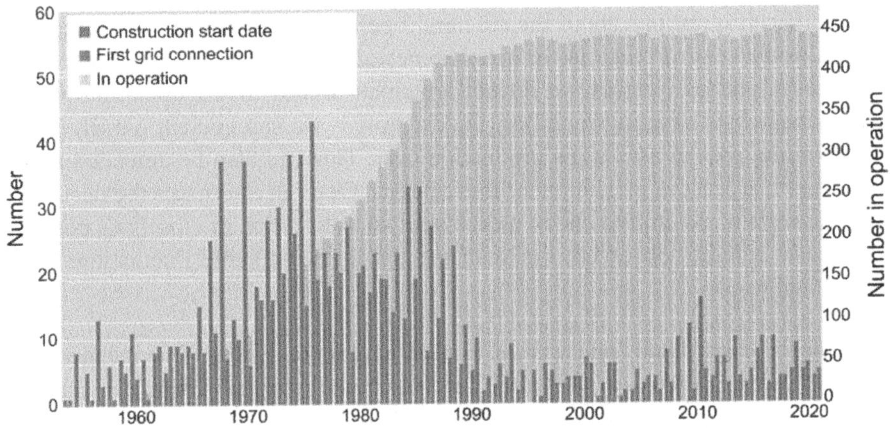

Figure 3.1. *History of the construction and connection of nuclear reactors (left scale) and evolution in the number of reactors in operation (right scale) around the world (source: PRIS database, IAEA). For a color version of this figure, see www.iste.co.uk/percebois/nuclear1.zip*

The current world fleet is largely made up of Generation II water reactors. Most of the world's fleet was built several decades ago (except for China and new "entrant" countries: see Figure 3.1). The current fleet is already, if not old, at least "mature" with an average age over 30 years, which explains the importance of the lifetime extension issue. The literature is quite rich in articles and books analyzing the construction costs of these reactors[3]. Two main questions pervade these studies:

3 Let us particularly mention the following: Moynet 1984; Parsons and Du 2003; Koomey and Hultman 2007; MIT 2009, 2011; Grubler 2010; Thomas 2010; Hogue 2012; Locatelli and Mancini 2012; Duquesnoy 2013; Haeseleer 2013; Berthélémy and Escobar 2015; Escobar and Lévêque 2015; Rothwell 2015; Lovering et al. 2016, 2017; Koomey et al. 2017.

(1) Is nuclear power becoming increasingly expensive, to the point of having to leave the market? (2) What are the different explanatory factors for the differences between countries and periods, and what do they tell us about the future?

3.2.2.1.1. Evolution of reactor costs in the United States

Figure 3.2 provides an overview of reactor costs for the United States. This type of study is constructed via data compilation, an equivalent scope, taking into account the monetary drift for each reactor and, if necessary, an exchange rate to determine these costs in a single currency.

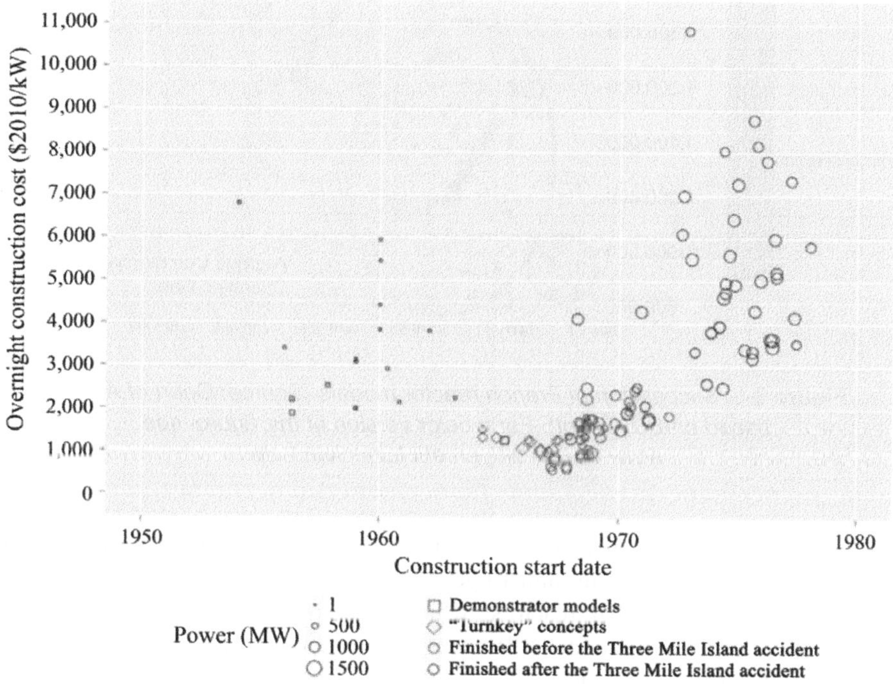

Figure 3.2. *Evolution of historical reactor construction costs in the United States (Lovering et al. 2016). For a color version of this figure, see www.iste.co.uk/percebois/nuclear1.zip*

3.2.2.1.2. Evolution of reactor costs in France

For France, the data referred to today are those compiled by the Court of Auditors based on EDF input. They are presented in Figure 3.3. The cost figures provided by the Court, deflated by the GDP price and fairly close to those analyzed by Grubler (2010), show an increasing trend in costs (the dotted curve drawn by the Court) of around 2%/year.

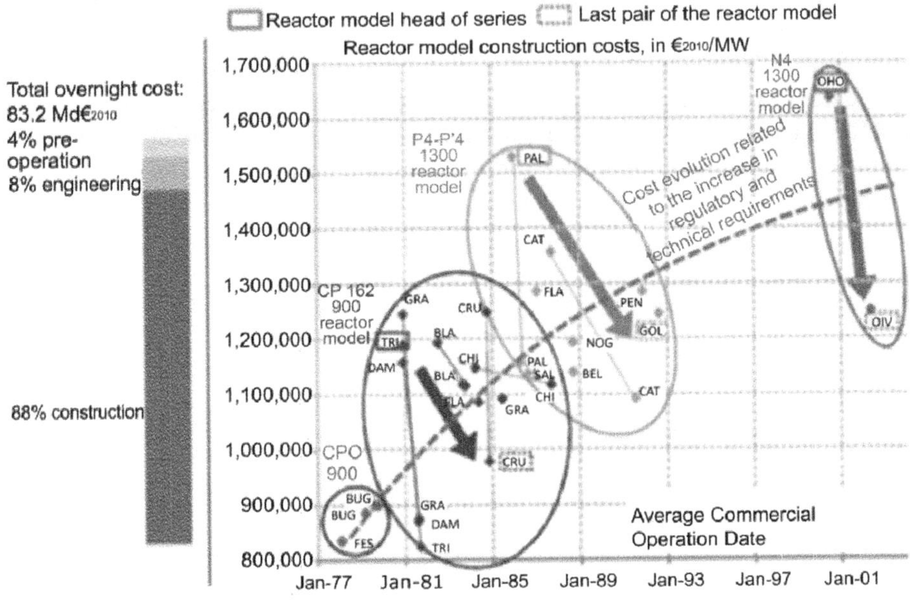

Figure 3.3. *Succession of French reactor models (Source: Court of Audit based on EDF input). For a color version of this figure, see www.iste.co.uk/percebois/nuclear1.zip*

The measure of this increasing trend in the real cost of reactors is quite sensitive to the low cost by Fessenheim and Bugey and to the higher construction costs of Generation II units (the N4 reactors, which were penalized by short tiers and lengthened construction schedules). It mainly depends on the deflator (price index) used. We will challenge this view.

3.2.2.1.3. Factors explaining reactor costs

Two main methods make it possible to explain the significant differences (factor of the order of 3 in cost/kW) between reactors and countries. The first one is to explain these costs via econometric regressions (Thurner et al. 2014; Berthélémy and Escobar 2015; Loving et al. 2016, 2017). The explanatory variables retained in these analyses are the number of reactors built, size, design, country's level of industrial development and construction schedule (which can in turn be endogenized). The second method relies on technological approaches (project cost figures by engineering teams) to reduce costs to a "standard reactor" level (Duquesnoy's (2013) approach). It is also possible to mix the two approaches. With a solid bibliographic background, we can summarize the main explanatory factors:

– **Type of reactor**: the different designs have specific costs influenced by the quantity of materials used. The physical volume of the reactor is different depending on the technology used: gas, water or molten metal. Similarly, the number of nuclear containment structures or the presence or absence of a core catcher play a significant role. Loving et al. (2016) propose an econometric approach combining types of reactors and construction schedules in function of design and "constructability", among other considerations.

– **Reactor power**: for a given concept, the effects of size (or scale) are important. Technological cost analyses enable their modeling thanks to the so-called Chilton formula. With standard size effects, for a double size, in a limited power range (if the concept can be extrapolated without difficulty), the unit cost can typically decrease by 13–25%. But conversely, large sizes can lengthen schedules and generate higher financial risks. Berthélémy and Escobar (2015) econometrically show that increasing the size of a reactor by 10% leads to a construction schedule 3% longer. However, it is the scale effect that has historically dominated and pressed to increase the size of reactors between 1950 and 2000. For a more comprehensive historical approach, see Locatelli and Mancini (2011).

– **Country effect**: the particularities of the various countries are extremely diverse, starting with the cost of labor, which is one of the major explanatory factors of the total cost. Let us also mention the industrial and labor organization, regulations and the organization of communications with the Safety Authority. Depending on the monetary parity choices made, these effects may have an influence that can reach a factor 2 between the extremes.

– **Construction of a pair of reactors**: it is considered that the second reactor in a pair costs around 10–15% less than the first one (SFEN 2018).

– **Program and series effect**: these (Joskow and Parsons 2009; Boccard 2014; Hansen and Percebois 2019) play a major role through standardization and risk reduction, the accumulation of experience, etc. France has successfully relied on this effect. Figure 3.3 shows the representation drawn by the Court of Auditor, with unit cost reductions within each level. The experience acquired makes it possible to rank reactor costs in a program. In Kee (2010), the sixth reactor in a series can reach a total cost almost 40% lower than the first unit. In addition, standardization contributes to reducing construction schedules (Berthélémy and Escobar 2015).

– **Site effect**: depending on land price, the amount of earthwork required, whether the site is already nuclearized or not and – above all – whether cooling is more or less easy (the need to build cooling towers), the cost of the site can amount to 10–25% of the total investment (NIRAB 2020), with significant variability (for example +10% for a site by the river).

– **Construction schedule**: this effect is highly significant and can, on its own, lead to a doubling of the cost. The shortest construction schedules were recorded in Japan and South Korea (less than 5 years), whereas the longest schedules are counted in decades (resuming construction after a work stoppage). Figure 3.4 illustrates the performances of six countries. Long delays generally have multiple consequences:

- increased total construction cost;

- loss of electricity production due to a delay in commissioning;

- a matching increase in financial costs;

- increased market risk. Berthélémy and Escobar (2015) explain that extending the construction schedule by 1 year can lead to an increase over 20% in the construction cost.

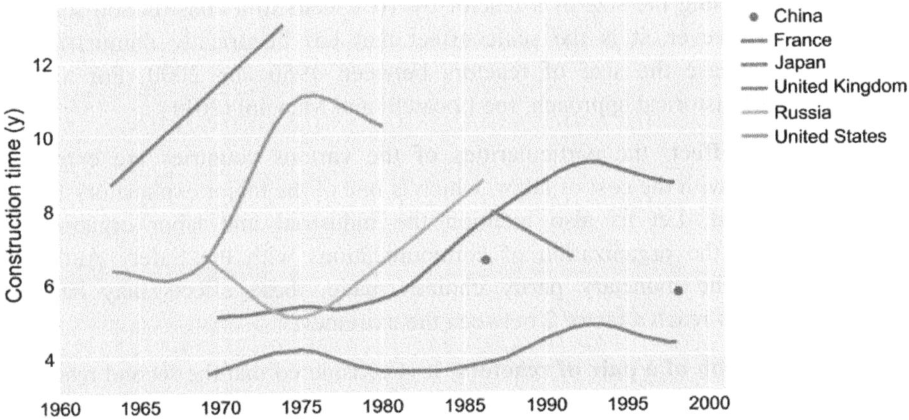

Figure 3.4. *Construction duration of "historic" reactors in six countries (source: PRIS database (IAEA)). For a color version of this figure, see www.iste.co.uk/percebois/nuclear1.zip*

3.2.2.2. *Other costs*

3.2.2.2.1. Operation and maintenance costs

These costs account for about a quarter of the full production cost. They mainly include operator salary costs and intermediate consumption (consumables, servicing and maintenance, miscellaneous assistance, administration, taxes and duties, insurance, etc.). Ultimately, they are made up of labor costs (related to the operator or subcontracting companies). They also include components such as insurance, communication costs and eventually, a share of the costs from the central services of electricity companies. It is customary to separate fuel costs, which have strong specificities (these are further discussed in section 3.2.2.2.2). A significant part of these costs is made up of the salaries of the reactors' former operators. Part of the scale effect of large reactors is embodied in the operation, bearing in mind that it takes about 400–500 people to operate a reactor over 1,000 MW, with little sensitivity to a larger size. Another type of cost is made up of equipment and works, which can partly be capitalized. For a full economic cost (FEC[4]) reasoning, the question that arises is that of the dynamics (temporal profile) of these costs and their recording (over time, or by taking into account the nature of the investment, with effects over many years). The Court of Auditors makes this type of choice explicit (Cour des comptes 2016). These costs include a significant fixed part and small variable part (around 20% of operating costs for a French PWR 1300).

In general, there are relatively no details about these costs at the international level and to our knowledge, there is no universal nomenclature for a thorough comparison among electricity suppliers. Data from electricity suppliers' accounts exist, of course, but their analytical breakdown is neither standardized nor public.

In the United States, there has been an increase in operating costs since the 2000s (peaking at more than $22/MWh in 2011), which have decreased by a few dollars over the last decade. In particular, this is due to the fact that, since 2003, the fleet in the United States has significantly evolved despite construction weaknesses:

– replacement of vessel heads for many reactors;

– replacement of steam generators;

– increase in production capacity;

– preparation for lifetime extension from 40 to 60 years with a raising in standards following the Fukushima accident;

– increased site protection.

The year 2012 marked the end of major investments in the fleet in terms of capabilities increase and lifetime extensions (from 40 to 60 years), which has resulted in a relative decrease in production costs. Nuclear power plants in the

4 The concept of full economic cost is presented in a former section.

United States compete against combined cycle gas power plants on the electricity market and have been forced to reduce costs in the face of the arrival of shale gas. A program was developed by electricity suppliers, *Delivering the Nuclear Promise*, with the goal of reducing production costs by 30% (including fuel and other components) below 40 $/MWh – the 2012 value – and reaching 28 $/MWh. In broad terms, the United States currently has an operating cost of around 15 $€_{2019}$/MWh (NEI 2020).

Other international data can be found in the reports by OECD/IEA+NEA (2015, 2020). They are of the same order, with lower values for China (by a factor of up to two) and significantly higher values for Japan (about 50% more than in Europe).

As already mentioned, the operating costs in France have been analyzed in detail by the Court of Auditors (2012, 2016). This shows that between 2006 and 2011 the situation had deteriorated. Then, EDF reacted, in particular by reorganizing the maintenance policy: better national and local coordination, creation of an operational management center and improvement in the management of equipment and stocks. Costs thus returned to more satisfactory values. The breakdown of French costs for 2013 is shown in Table 3.1.

	Labor	External consumption	Taxes and duties	Central functions	Total
Amount in $€_{2013}$/MWh	7	5.6	3	2.3	17.9

Table 3.1. *Operating costs (excluding fuel) of French reactors over the period 2008–2013 (Cour des comptes 2012, 2014, 2016)*

In total, they amount to almost 20 $€_{2020}$/MWh. These are average costs for a fleet undergoing major transformations. Thus, the French fleet is currently still subject to potential factors of increase: salaries, procedures, impact of the major refit on operations, impact resulting from COVID-19, etc. Conversely, an analysis conducted for a new reactor shows significantly lower results.

At a global scale, on average, these costs represent about one-third of the accounting costs of the nuclear electricity production, with a non-negligible standard deviation (about 10 points according to D'Haeseleer (2013)).

Operating costs are an essential part of the *cash* costs and, in this sense, they play a decisive role in the future of reactor operation. In the case of the United States, in its reference scenario, the Energy Information Administration (EIA) calculated that a 20% reduction in operating costs could eventually reduce the final shutdowns of reactors due to loss of competitiveness by almost 5 GW or 18% in the next 30 years (Scott 2018). But these costs also reflect essential actions to ensure proper operation, and in particular, a high reactor availability. They can be favorable to safety as well as economically profitable. Among the levers for improvement implemented on a large scale, let us mention the intensification of training, modeling (including the "digital twins" of reactors) and predictive maintenance.

3.2.2.2.2. Fuel costs

The costs of the nuclear fuel cycle are well known and have given rise to an abundant bibliography. They are discussed in detail in the dedicated chapter of this book. They include the following components:

– Cycle front-end: uranium conversion, enrichment, manufacture.

– Cycle back-end: storage, transport, treatment, (recycling[5]), conditioning and long-term waste management (in the case of countries having opted for the so-called "closed" cycle) or storage, transport, conditioning and long-term waste management (in the case of countries having opted for the so-called "open" cycle).

From an accounting point of view, these costs have several particularities. On the one hand, an important portion is fixed, insofar as the shutdown programs for unloading and reloading reactors do not exclusively depend on the fuel exhaustion level (combustion rate). On the other hand, they both involve cash costs (for the purchase of materials and services), financial costs (for the financing of stocks of materials and services, starting with uranium (Cour des comptes 2014) and non- cash costs (provisions for future charges), which often have the particularity of being covered by dedicated financial assets. In France, the assets in question currently cover all expected future expenditure for decommissioning, recovery and conditioning of old waste and corresponding waste storage. It is via these dedicated funds that the cost of CIGEO storage is and will be fully funded.

Regarding the calculation methodology, the OECD/NEA have extensively developed tools making it possible to associate the cost of fuel with the cost per kWh (OECD/NEA 1994, 2013). Note however that the general hypothesis of fuel cost separability is far from trivial. It allows this cost to be considered as an instantaneous expense during the year of electricity production: to do this, the fuel

5 The recycling component is in brackets because it can be a credit or a debit, depending on the economic conditions and the methodology used (Devezeaux de Lavergne and Durret 1995).

cost has to be calculated with the same discount rate as the cost per kWh, something which we recommend. In the case of the so-called "open" cycle, there is no other subtlety. On the other hand, in the case of the "closed" cycle, which involves recycling, the hypothesis for "dissociating" 1 kg of fuel from the future of recycled materials requires giving a value to these materials. It is clear that this value depends on the hypotheses made regarding the future of the nuclear fleet, and will be different in the event of the development of Generation IV reactors, or a fleet shutdown at the end of the operating period. A comprehensive method is described in De Roo and Parsons (2011) and outlined in Chapter 2. These amounts also depend on the hypotheses made as to the costs and time frame for waste storage (Devezeaux et al. 2020a). All things considered, the economic analyses show that the open cycle might be a little less expensive than the cycles involving recycling: the difference would however be of the order of 1% on the production cost. The multi-criteria analyses of the different cycles, which take into account material balances and environmental impacts, generally show more advantages for the closed cycle (Poinssot et al. 2014).

Another point of historical interest is the dependence of fuel cost on the combustion rate. This rate's increase, from the value of the design of the first 900 MW reactors (33 GWd/t) to current combustion rates (up to 50 GWd/t) has enabled a significant reduction in fuel costs, bringing the operation of current reactors to a zone close to the minimum cost. Debès (2002) mentions a decrease in the unit cost of fuels, in €/MWh, of almost 25% since water reactors started operating in France via the increase in combustion rates.

For the French operator, on the basis of the complete documentation made public in Cour des comptes (2014), the front-end costs for the EDF cycle amounted to €5–6_{2020}/MWh during the 2010s. This order of magnitude remains valid. Back-end expenditure can be estimated based on the releases of provisions for fuels and waste appearing in EDF's reference document: over €3_{2020}/MWh. However, it should be understood that this is an approximate *cash* cos over the year, which is related to production. As soon as the economic question of the cost of a duly identified MWh arises, one should bear in mind that the back-end costs of a given year are actually related to fuels whose energy has typically been produced over 10 years ago. If one chooses to update these amounts, they have to be divided by two or more. The report from the Court of Auditors (2014) also stresses that doubling the cost of waste management would only increase the ex ante nuclear average cost by about 1% (LCOE-type).

For other countries, the United States currently display a cycle cost of around €5_{2020}/MWh (NEI 2020). The values provided by OECD/IEA+NEA (2020) are comparable and fairly homogeneous, which reflects the strong internationalization of cycle component markets. In total, the accounting costs (*cash out*) of fuel from a nuclear power plant in OECD countries amounts to around €5–8$_{2020}$/MWh. This fuel cost represents approximately 15–20% of the accounting production cost of nuclear electricity.

3.2.2.2.3. Financial costs

In practice, it is quite elusive to explicitly relate financial costs to the operation of one or more reactors. This type of approach regarding a historic fleet is at the heart of Champsaur-type methods for ARENH or FECs. The method comes down to translating the cost of capital into accounts. There is no homogeneity in the financial costs of nuclear electricity, as situations dramatically differ from one company to another. There is often a confusion with the interests during construction calculated via the LCOE method for new projects. This choice is not justified: the capital is not always borrowed in full and the interest rates incurred are not discount rates, especially if public rates are used. The order of magnitude of these accounting financial costs for historical nuclear power oscillates between a few percents and 15% of the electricity production accounting cost.

3.2.2.2.4. R&D costs

Internationally, R&D costs for different energies have mainly been compiled by the OECD/IEA. Leprince and Popiolek (2016) describe the composition of these costs and compare various situations among several large countries. These costs constitute a relatively modest component for the operator, knowing that part of the R&D is carried out with public funds. Part of the overall expenditure passes through the electricity bill and the other part through the State budget. In total, the countries from the International Energy Agency (IEA) currently spend around €4bn$_{2018}$/year in nuclear R&D. Compared to the nuclear production of the OECD, the expenditure amounts to nearly €4$_{2020}$/MWh.

As for France, the Court of Auditors (2012) calculated that the R&D funds destined to water reactors amounted to €43 billion$_{2010}$, over the period 1957–2010. Annualized and then compared to the annual energy produced (a little less than 400 TWh, at present), this amount is slightly lower than the previous one, around €2.5$_{2020}$/MWh. One has to add about €0.5$_{2020}$/MWh for the development of Generation IV reactors. France invests proportionally less in nuclear R&D than the OECD average. This result is not paradoxical if we consider the size of the stock: this is also a scale effect. About half of the total is borne by industrial partners. Thus, the public nuclear "subsidy" via R&D amounts to around 3–5% of the full production cost. Many academic studies also support the thesis that it is socially effective to support R&D devoted to long-term programs, especially if these involve high technology and generate spillovers. This applies not only to nuclear, but also to renewables. This mechanism is found in the "Rohmer" models.

These expenditures have played a major role in the performance of the world's nuclear fleet, in several directions: continuous safety improvement, design amelioration (technological efficiency, for example, of yields), operation optimization, power increase (decalibration) for certain reactors, increase in combustion rates, spent fuel and waste management recycling, etc.

R&D costs represent approximately 5% of the accounting production cost of nuclear electricity in the electricity supplier's accounts.

3.2.2.2.5. Decommissioning costs

The literature on the subject is vast and is based on two distinct sources: operations already carried out (several dozen worldwide, including the pressurized water reactor [PWR] reactor of Chooz-A by EDF) and the engineering calculations for the future, which are themselves constantly updated on the actual data.

In this chapter, we will address the decommissioning question extensively, which is also the subject of the Appendix in, Volume 2. An order of magnitude of 15% of the construction cost can be adopted for serial decommissioning: this is the value calculated by EDF in its exercise for the Dampierre site. On this basis, at the end of 2019, the group's decommissioning provisions amounted to €32 billion.

From an economic point of view, these are charges observed well after the production of electricity (as it happens for the long-term management of waste). In practice, electricity suppliers make accounting provisions (in the income statement) and also generally supply dedicated funds to cover these expenses in the long-term (see the section below which specifies these mechanisms). As noted by the Court of Auditors (2012, 2020b), the impact on production costs is still modest, since these amounts are significantly lower than the initial investment and the expense is remote. It typically amounts to a few percents, regardless of the type of cost considered. In addition, as can be seen on the basis of this modest value, errors or uncertainties on components that are subject to significant contingencies (regulations, costs of unit operations, schedules) are not able to alter the cost of nuclear production.

3.2.2.3. *Reactor lifetime extension*

This section no longer deals with a cost component, but with the application of economic methods to the choice of whether or not to extend the lifetime of a reactor.

This question arises essentially in relation to safety, the authority in charge issuing operating permits. The authorization may be issued on a conditional basis, subject to works. The question is whether the cost of such work is economically justified. Otherwise, the reactor ends its operation and is to be dismantled.

In France, a first response to this type of question is provided by the application of the FEC method by the Court of Auditors. The complete calculation has been notably formalized by IAEA (2018) and OECD/NEA (2021b). The decision to pursue an operating extension program mainly depends on a comparison of the costs of operating the plant's production – including maintenance, fuel and the investments necessary for the extension – with the cost (this time including construction) of alternative sources of electricity (to be developed in the event of reactor shutdown). The costs considered are disbursements. They are updated with a cost of capital and take into account the risks associated with each decision. Time plays a central role, due to the delay in certain operations (delaying decommissioning can generate a non-negligible economic value).

At the international level, the studies by the OECD/NEA show that the analyses results almost totally agree on an exploitation extension, because refurbishing costs are relatively lower compared to the investment cost in new production means, in view of a growing or constant demand. The range of refurbishing costs, estimated in 2012 by the NEA, was between 500 and just over 1,000 $US\$_{2010}$/kWe. The so-called "post-Fukushima" provisions are included in these values and account for a share of 10–17%. The CEA/I-tésé studies reported by Percebois et al. (2012) show that it is from 2 billion euros/unit that things could change and that due to "micro-economic" reasons at the reactor's level, EDF could vote on shutting down part of the fleet. But this point has not been reached, yet. A recent estimate of the cost of the French major refit amounts to €50 billion or approximately €1 billion/reactor. A last question, finally, is that of the desirable operating life of reactors, the major refit aiming for a duration of 50 years, with no guarantee beyond that. Currently, the analyses by EDF or RTE (2021) cover a period that does not exceed 60 years. When the time comes, it will also be necessary to question such a limit in France.

Symmetrically, the underlying reason for shutting down a reactor could be economic. For example, D'Haeseleer (2013) refers to the case where operating costs are too high compared to other production means, or to the market price (especially when the latter is regulated). In such a case, it is possible for the operator to prefer shutting down the reactor, even if its technical and regulatory condition enabled it to pursue the exploitation. In particular, this type of early shutdown occurred in May 2013 in Wisconsin, where the Kewaunee Nuclear Generating Station was shut down, even though it was granted an operating extension license by the NRC until

2033. The main reason for this was the low level of electricity prices in this state, driven down by the price of shale gas.

All in all, only rare situations such as the above-mentioned economically justify shutting down a nuclear reactor whose operation has been authorized by the Safety Authority.

3.2.2.4. *Historic nuclear competitiveness*

Since its origins, nuclear power has quickly diversified from a technological viewpoint (1950s and 1960s) and spread to the major industrial countries, with diverse economic realities. In its very beginnings, its profitability was low, because the small units did not benefit from a sufficient size effect. In the 1970s, industrial standards converged on water reactors, which were notably capable of high power[6]. Nuclear power had then reached technical and economic maturity.

Since the 1970s, costs have increased, essentially under the effect of increases in input costs: mainly labor and materials (in the United States, the increase in the 1970s was brutal, following changes in regulations). We have also seen that the impact on the costs of ever-increasing safety standards was difficult to quantify, at least with the Generation II reactors. All things considered, the competitiveness of existing nuclear power has been generally maintained over the last decades. In France, the FEC is around €60/MWh, including the fleet's refurbishing costs.

In addition, both the work of the SFEN (2017) and the WNA (2020) show that the annual average *cash* costs of nuclear power are stable[7] and competitive. These costs have remained low (around €30 to €40/MWh in France) ensuring the interest of operating the fleet almost everywhere, even taking into account the costs involved for lifetime extensions while respecting the highest safety levels. The short-term marginal costs are extremely low (around €10/MWh), which ensures a regular call for production from operating reactors, even if the production coefficient erodes with the arrival of renewables. This competitive advantage could increase if externalities were taken into account.

Conversely, adverse events and trends have emerged. Some have been visible, such as the Chernobyl and Fukushima accidents. The latter seriously slowed down the resumption of construction that had been announced at the time (the "renaissance" of world nuclear power or "nuclear revival").

6 In their vast majority, gas reactors have a low power density, which in practice did not make it possible to reach powers of the order of 1,000 MW or over with that technology.

7 A high share of national added value and low exposure to the price of uranium.

Other developments were much less perceived, such as the undeniable loss of competences in engineering and in the sector, due to the lack of a large-scale construction programs. In Europe in particular, operating costs have slowly drifted, as it has been repeatedly pointed out by the Court of Auditors in the case of France. At the same time, the availability coefficient has eroded. Over the past 10 years, these signals have been identified, and major plans have been put in place to improve the situation.

The big question that now arises, in relation to the future place of this energy in the global electricity mix, especially in the perspective of the fight against climate change, is the performance of new nuclear power.

3.3. Costs of nuclear electricity at the power station terminal

It has been shown that the construction of reactors at a global scale (essentially in the OECD) has followed a marked cycle, with a sharp decrease in activity in the years 1990–2000, when grid connections had fallen to 3–4 per year in the world. Since the end of the previous century, intense design work for new reactors has been carried out, with a new safety perspective.

In Europe, this was initially framed by specifications for electricity suppliers, called European Utilities Requirements (EUR), which was precisely at the origin of the EPR-line concept. These works, which had significantly evolved in comparison with previous standards, were grouped under the Generation III term, together with their American and now Korean and Chinese counterparts. The technical goals targeted by all these concepts can be summarized as follows:

– probability of core meltdown <10–5 per year and use of probabilistic studies during design;

– in the event of a serious accident, minimal impact on the population by maintaining containment;

– reinforced protection against external aggressions (natural and human);

– availability >90% over at least 20 years;

– 60-year operating life since design.

The purpose of this section is to briefly describe the cost bases, then the increases encountered and the extension of the construction schedules for certain first series, to explain the lessons learned and to specify the expectations in terms of costs for this new generation. Apart from that, we will discuss the question of the cost for other reactors (small and medium reactor [SMR] and fast neutron reactor [FNR]) and outline some perspectives in the medium and long-term.

3.3.1. *Construction costs of 3G water reactors*

3.3.1.1. *The difficult experiences of building head of series Generation III reactors in Europe and the United States*

At the global level, toward the end of the 2010s, the market appeared to be partly revived with an annual flow of nearly 10 nuclear power reactors commissioned (almost exclusively Generation II), something which had not occurred for two decades. In 2020, 50 reactors were under construction. A growing share of these new reactors is now Generation III, and the heads of series for the main concepts are now underway. Thus, the first EPRs were differentiated: Framatome (France), AP1000 Westinghouse (Japan-US), VVER 1200 Rosatom (Russia) and Hualong 1 CGN-CNNC partnership (China). Here, all of these concepts will be classified under the Generation III, without bringing up specialist discussions on the exact boundary of the classification. We could add the APR KHNP (South Korea), but this concept does not reunite all the characteristics of Generation III reactors.

Russia has been working for 8 years on the construction of its first Generation III reactor (since 2016, in Novovoronezh). In this country, the construction programs for Generation II reactors was not interrupted as in Europe and the United States. This is also the case for South Korea, which met its deadlines for its AP1400 "transition" model at Shin-Kori. Then, Taishan 1 was the first EPR by Framatome to enter service worldwide in 2018, soon to be followed by Taishan 2. China's first reactor, Hualong-1, entered into service in November 2020, after just over 5 years of construction. On the other hand, the two European EPRs of Flamanville and Olkiluoto (Finland) suffered significant delays and were still not operational in mid-2021 (over 15 years of work in total). In the United States, ongoing projects with the AP1000 model (Vogtle 3 & 4 and Summer 2 & 3) also suffered delays and additional costs to such an extent that the construction of Summer 2 & 3 was abandoned: in fact, the US electricity market offers little prospect in the short-term, with a low price driven by the price of unconventional gas. As with the EPR, the Chinese AP1000 units began their activity on (or near) the schedule initially set (the delay was mainly due to the lack of feedback expected from the United States).

	Reactor	Country	Capacity (MWe)	Construction start date	Initially scheduled starting date	Starting date	Initial construction cost (€/kWe)	Revised construction cost (€/kWe)
EPR	Flamanville 3	France	1,600	Dec. 07	2012	In progress	1,800	>7,600
	Olkiluoto 3	Finland	1,630	Aug. 05	2009	In progress	1,700	5,700
	Taishan 1	China	1,660	Oct. 09	2014	2018	1,700	3,200
	Taishan 2		1,660	Apr. 10	2015	2019		
	Hinkley P. C1	UK	1,630	Dec. 18	2025	In progress	6,400	7,500
	Hinkley P. C2			Dec. 19	2026			
AP1000	Summer 2	United States	1,117	March. 13	2017	Canceled	3,700	Not available
	Summer 3		1,117	Nov. 13	2018	Canceled		
	Vogtle 3		1,117	March. 13	2017	In progress	~1,000	~9,000
	Vogtle 4		1,117	Nov. 13	2018	In progress		
	Haiyang 1	China	1,000	Sept. 09	2014	2018	2,200	Not available
	Haiyang 2		1,000	Jun. 10	2015	2018		
	Sanmen 1		1,000	Apr. 09	2015	2018	2,200	2,600
	Sanmen 2		1,000	Dec. 09	2016	2019		

	Reactor	Country	Capacity (MWe)	Construction start date	Initially scheduled starting date	Starting date	Initial construction cost (€/kWe)	Revised construction cost (€/kWe)
VVER 1200	Novovoronezh 2.1	Russia	1,114	Jun. 08	2015	2016	3,200	Not available
	Novovoronezh 2.2		1,114	Jul. 09	2016	2019		
	Leningrad 2.1		1,085	Oct. 08	2013	2018	2,700	Not available
	Leningrad 2.2		1,085	Apr. 10	2016	2020		
	Ostrovets 1	Belarus	1,194	Nov. 13	Not available	2020	Not available	Not available
	Ostrovets 2		1,194	Apr. 14	Not available	In progress		
	Akkuyu 1-4	Turkey	1,114	Apr. 18	2023	In progress	Not available	Not available
	Rooppur 1 & 2	Bangladesh	1,180	Nov. 17	2023	In progress	Not available	
HUALONG 1	Fuqing 5 & 6	China	1,000	May. 15	2020	2020 (No. 5)	2,300	2,900
	Fangchenggang 3.4		1,090	Dec. 15	2020	In progress		
	Zhanhzhou 1&2		1,126	Oct. 19	2024 & 2025	In progress	Not available	Not available
	Taipingling 1&2		1,126	Dec. 19	2025 & 2026	In progress	2,500	Not available
	Karachi 2 & 3	Pakistan	1,014	Aug. 15	2021	2021 (No. 2)	4,200	Not available

Table 3.2. *Global census of Generation III reactors connected or under construction at the beginning of 2021 (source: Compilation by the author of information published by the WNA, the OECD/NEA and others). The cost values indicated are not homogeneous and their comparison is only indicative*

Table 3.2 lists the main ongoing projects for the new Generation III models, with their construction schedule and available cost information. The number of reactors of this new generation under construction is already considerable: several dozen. Note that while many other concepts have been developed, they are not currently under construction (such as Framatome's Atmea 1).

There is a large dissymmetry between the countries which have pursued a construction program at a good pace (China on top of the list) and the others (certain OECD countries), affecting not only the construction schedule, but primarily the cost of the first Generation III reactors. This has been summarized (OECD/NEA 2020) in the following graphical form (Figure 3.5), with a factor of around 3 in comparison with the "old world".

This figure also makes it possible to appreciate the cost of Generation III reactors compared to that of "historical" reactors. One can observe that today's Asian costs are close to the costs of the early 1970s in OECD countries. Costs in Asia benefit from labor costs which are still lower than those of the OECD, as well as from less restrictive worker protection rules. But it is above all the quality of the organization, largely favored by a regular production rate, which explains the current gap between large areas. These figures greatly nuance the articles that describe a "jump" between the average costs of Generation II and III reactors, allegedly more expensive from a structural point of view.

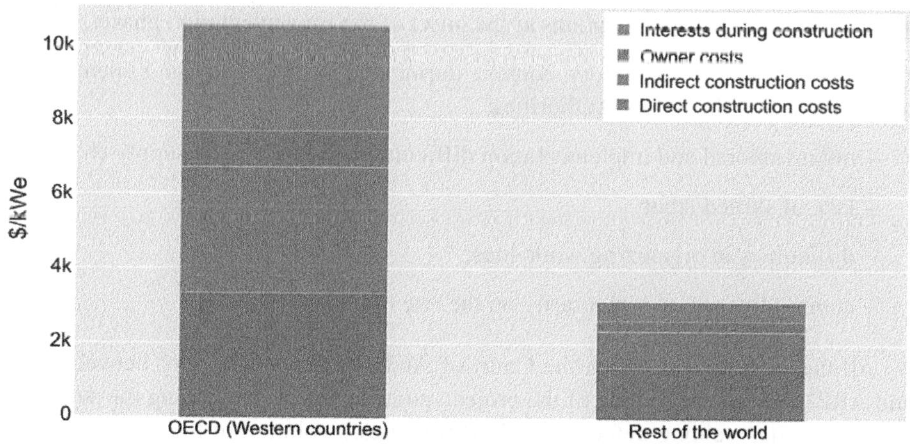

Figure 3.5. *Comparison of the construction costs of Generation III reactors, depending on the region of construction (source: OECD/NEA). For a color version of this figure, see www.iste.co.uk/percebois/nuclear1.zip*

Such a counterperformance of "Western OECD" countries deserves to be explained. We recommend reading OECD/NEA (2020) on the analysis of these derivatives. But, beyond that, these profound difficulties must be analyzed in anticipation of the future: if American and European costs are not reduced by a factor of at least two, the revival of nuclear power in these continents will not take place. We already have the tools to decipher the main cost factors: construction schedule, site by the sea or not, construction in pairs or not, etc.

The largest drift is observed for the two European EPRs and for the US AP-1000, bearing in mind that these were the first two Generation III concepts to have been engaged. Thus, the constructions of Olkiluoto 3 and Flamanville 3 began in a doubly risky context:

– "head of series" effect of the new EPR reactor concept, incorporating the requirements listed above, unit power never before attained;

– absence of construction of a nuclear power plant in France and Europe for over a decade.

Apart from the "head of series" effects, the causes for the drifts of the first EPRs have been analyzed in Folz (2019) and Cour des comptes (2020a) in great detail. The main factors are mentioned below, quoting the analysis by SFEN (2018):

– lack of initial maturity of the concept (in the industrial sense) and too incomplete technical specifications at the onset of the implementation phase;

– changes in the regulatory context during the project and (in Finland) new setting regarding the Safety Authorities;

– organizational and implementation difficulties throughout the supply chain;

– lack of skilled labor;

– difficulties in organizing work sites;

– commodity prices, temporarily on the rise in the 2010s.

All these factors, to which the Court of Auditors adds the rivalry between EDF and AREVA at the beginning of the project, contributed to lengthening the deadlines (rejected parts, welding defects, supply difficulties, evolution of the concept during the construction, etc.). The survey carried out by the Boston Consulting Group for the SFEN in 2019 highlights the great difficulty finding qualified labor (welding, in particular).

The construction schedules for certain heads of series extended from 5 to 15 years in order of magnitude. Based on the results mentioned by Berthélémy and Escobar (2015), such a deadline extension can lead to an increase in the construction cost by a factor between 3 and 6. Thus, the duration of works is largely responsible for the overall drift. Other causes have also contributed, from the outset of the projects, to the high cost of these reactors: single reactor (no pair effect) and extensive excavation works for Flamanville 3, in particular.

EDF assessed the impact of a 3-year delay for entering into service of an EPR (set on Flamanville) in terms of return time. This construction was behind schedule by over 10 years.

As recalled by the University of Chicago (2011), beyond production, the uncertainty generated by the lack of vision in the orders also had serious consequences. This last cause of increased costs, combined with changes in contractual rules, is also included in OECD/NEA (2015). According to the agency, a major reason for the increases in construction costs was due to new contract conditions throughout the supply chain, mostly with fixed or firm prices. The turnkey contractor could think of removing the drift risk for his costs. But these contracts were concluded at the cost of a significant risk allowance, due to the caution of industrial suppliers. At the forefront, they also tried to hedge themselves by negotiating similar contracts with their own suppliers, and so on, which ended up creating the stacking effect of contingency provisions.

3.3.1.2. *A first step in cost reduction*

Regarding EPRs, the much better Chinese performance was only due to the location of Taishan projects in China. It also benefited from European setbacks. Areva, then the new Framatome, had been working since 2009 to gather feedback from European projects "by the water". The Taishan reactors thus benefited from this feedback. The following examples illustrate the benefit provided by the European feedback:

– the reduction of the construction schedule between the first concrete facility and the installation of the dome (24 months instead of 47) between Olkiluoto 3 and Taishan;

– the significant reduction in the number of engineering hours, and the associated costs, on the nuclear boiler (–60% between Olkiluoto 3 and Taishan);

– the significant reduction (–40%) in the manufacturing time of large components because of the improvement of production processes (for example: one year reduction in the manufacturing time for the steam generator in Taishan);

– the reliability of the construction schedule because of the reduction of supply times with suppliers (reduction of times by 65% on average between Olkiluoto 3 and Taishan).

In addition, the number of hours spent by the engineering teams was 60% lower for Taishan than for Olkiluoto 3.

At the international level, the work in OECD/NEA (2020) was entirely devoted to cost reductions, and many articles and communications also focused on this subject. The analysis (EPRI 2011), structured in terms of curve of learning, remains fully valid. We reproduce it in Figure 3.6.

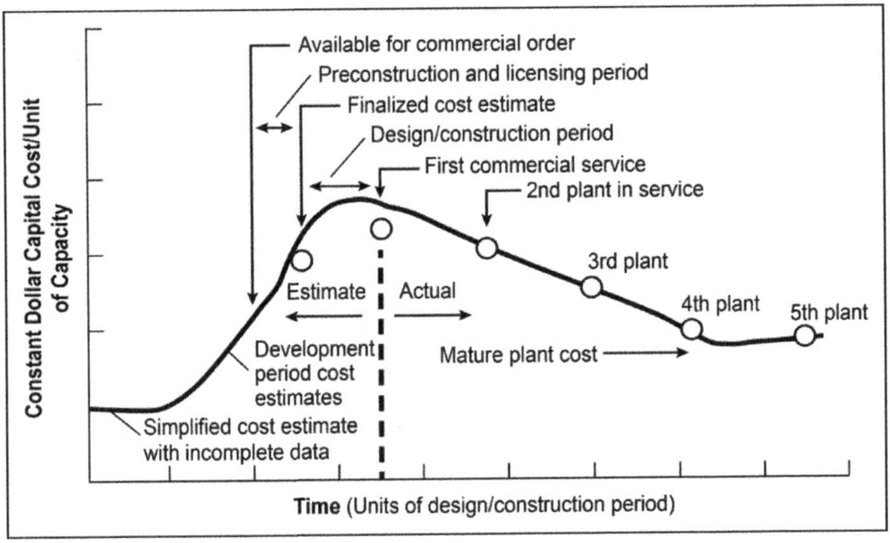

Figure 3.6. *Learning curve of reactor costs for a new concept (source: EPRI 2011)*

The learning curve describes the evolution of the costs of a reactor design in a stylized way. It incorporates the experience of the last decades and is also based on the characteristic curves of other industrial sectors, as shown by SFEN (2018). The first part of the graph is characteristic of a phase where the complexity of the design increases, the constraints of all kinds being gradually taken into account. These phenomena are known and the means to control them are not new (e.g. the implementation of integrated design platforms), bringing together the main trades concerned, including subcontractors: manufacturing equipment, assembly, construction, testing, operation, etc. But there is no half-baked recipe in these areas: the development of a "perfect" organization would result in excessive gestation times. Often, between the first cost estimates of the sketches and the costs of a finalized project, it is common for the amounts to increase by several tens of percents. One of the factors weighing on the costs of the heads of series is the simple amortization of engineering on a small number of reactors (unless one has a high order book from the beginning). Uncertainties are also greater at the start of the cycle. As we have seen, margin behavior within the supply chain can provoke significant costs.

These phenomena explain the passage of the learning curve by a maximum cost/kW corresponding to the phase of effective construction of the heads of series. Then, a significant number of effects come into play, which tend to reduce costs. They are explained in the following.

3.3.1.3. *An example of a reoptimized reactor concept: the EPR2*

One of the ways to limit the costs of Generation III reactor series involves modifying the reactor concept, after feedback from the first constructions. Framatome's goal since 2015 has been to simplify and "lighten" the reactor, but also to make it easier to construct, for example in terms of modularity, prefabrication or accuracy in the design and position of fixing plates equipment. The development of this new concept is also accompanied by an in-depth reflection on the methods and tools (digital in particular) to reinforce the efficiency of engineering teams. Let us mention the following features:

A boiler with the power level of the latest EPRs (thermal power of around 4,600 MW), enabling the best "reuse" of the EPR's primary equipment.

– A single-walled containment structure, but of increased thickness, with a liner providing both containment of radioactive materials in the event of an accident and protection against malicious external attacks. A "three-train" (no longer four) emergency back-up architecture. This choice aims to simplify the design.

– Optimization and simplification by integrating feedback (including the construction process in such a way as to generate significant savings in costs and worksite delays), reduction in purchasing costs, etc. This approach acknowledges feedback from suppliers and particularly focuses on:

- the simplification of building structures (sails alignment, removal of singular points);

- the optimization of civil engineering reinforcement rates;

- the optimization of electromechanical assembly sequences by relying more on prefabrication and reducing the number of hydraulic tests;

- equipment standardization via the use of catalogs based on industrial standards;

- better control of the control-order volume by reducing the amount of data transmitted by the equipment;

- the development of system engineering, applicable to complex projects.

Cost reduction via design enhancement is considerable. The study of SFEN (2017) refers to a 30% reduction in the overnight cost of the EPR2, compared to what was expected for Flamanville's EPR, at an equivalent site. But it is necessary to consider the actions described above as a coordinated whole, which can only succeed in the perspective of an industrial program: now it is the descent on the right part of the learning curve that starts, made possible thanks to a better design.

3.3.1.4. *Switching from a head of series to an industrial series: impact on costs*

The learning effect incorporates progress over time. Apart from working on the concept, we have detected a series effect in many countries resulting from the number of reactors built, which reflects the fact that the average specific investment cost (€/kW) of a series number of standardized units is lower than that of a tranche with the same characteristics but designed and produced in a separate manner. These are the gains acquired during the implementation of the programs. This effect is mainly provoked by industrial causes, but not only those. Not only does the progress of a program include securing the regulatory environment for the different tranches, but also progress and innovations in codes, standards and processes. In economic terms, the intention is to reduce transaction costs via a proper organization. This virtuous approach is described in OECD/NEA (2019), with the image of two "waves" after the head of series: a similar first stage, for the EPR, paired with the site under construction at Hinkley Point (UK), then a second stage which fully benefits from the maturity of the sector.

The predominant series effect is of an industrial nature. In our analysis published by the SFEN and quoted above, we have identified three distinct components: the program effect, the productivity effect and the pace of works (regularity and annual flow of orders to the supply chain).

3.3.1.4.1. Program effect

This effect originates in the action of the turnkey contractor (the industrial actor who pilots the project and supervises the construction of the reactor, for example, EDF in France). The uniqueness while carrying out the studies, developments, qualifications and tests of equipment valid for an entire level constitutes the program effect (or tier effect). The costs corresponding to the program's development are independent from the number of tranches involved and are not sensitive to unit size (power). They are all the better amortized as the number of units in the program increases. However, they are strongly influenced by the degree of innovation and complexity introduced into the reactor concept of the new series. In France, EDF wishes to build six EPR2 reactors from the mid-2020s. The decision by President E. Macron, at the end of 2021, to launch such an EPR2 program is intended to let this effect come into play in the coming decades.

3.3.1.4.2. Productivity effect

This effect mainly occurs at the level of the supply chain. It results in the repercussion of supplier productivity gains on their selling prices. It is highly dependent on the visibility given to suppliers via a firm order for a series of identical and standardized materials. This visibility makes it possible to set up an optimized planning and use of supplies, production tools and skills (businesses). It also makes it possible to gradually reduce costs during the program via the continuous improvement processes accompanying it.

3.3.1.4.3. Pace effect

The pace effect concerns all actors. It is closely related to the program effect. It reflects the fact that the beneficial effect of the quantity of reactors built is a function of the associated commitment rate, within the range authorized by the existing industrial potential. It translates the industry's load factor. For this effect to be noticeable, the number of tranches involved must make it possible to maintain a minimum continuous load, both in terms of studies and equipment manufacture. The continuity of commitments is an essential parameter

A volume effect is directly associated with the pace of construction. By definition, fixed costs (station amortization, skills maintenance) are important in the event of low annual flows. For example, while the use of IT has increased sharply over the past 20 years (with the associated productivity gains), the number of IT specialists in the operating teams has generally remained fixed. This effect comes into play as soon as the period (for which the frequency of orders is known) appears to have a solid basis for investing in equipment, training, recruitment, quality, etc., and in a long-term perspective, for launching specific R&D actions. In the nuclear industry, the characteristic time frame to achieve such rate gains typically takes decades.

In total, the series effect (resulting from the combination of the above-mentioned effects) is reinforced whenever it is possible to build several tranches of the same nature at the same site. Thus, site development works are pooled, as are certain works and ancillary buildings (the pair effect already described).

As an illustration of these effects, EDF estimates that it can reduce the construction cost of its two EPR projects by 20% at Sizewell C by transposing elements from the Hinkley Point project. For example, EDF believes it can dispense with most of the £2 billion (€2.2 billion) needed for pre-construction work at Hinkley Point C by hiring subcontractors currently working on site and using equipment which has already gone through the certification process required for use on another nuclear site.

For a country with a nuclear system, these different effects are entirely triggered by national programs. But they are also partly activated by export programs. Depending on the case, the value of exports when building a reactor abroad can typically represent 30–70% of the total construction overnight cost. In addition, major programs can stimulate other sectors with competing concepts: this is for example the case of VVER or Hualong that contains a share of equipment produced in France (SFEN 2018). However, the construction outside the country or zone of "origin" of a concept cannot suffice on its own to consolidate a complete industrial sector. On the one hand, the guarantee of stability is lower, in the eyes of investors, if only exports come into play, without a national base. On the other hand, the trades mobilized by civil engineering are the first to obey local logics, despite the fact that labor is largely fluid at an international scale. Finally, the purchasing countries seek to locate as many other trades and suppliers as possible in their country, so that the supply chains must be reconstituted for each case and are not easily transposable.

3.3.1.5. *Costs of future Generation III reactors*

In the 2020s, the Generation III concepts will reach their maturity. The Chinese and Russian reactors have already largely entered the industrial phase, benefiting from a continuous construction pace. For European, American or Japanese concepts, this phase is beginning. It would also be more accurate to reason in terms of construction area, not barely in terms of concept: take the successful experience of EPRs and AP1000 in China, while projects in Europe continue accumulating additional costs and delays.

The French example reveals that there are many ways to reduce costs, and some of them have already borne fruit. Still, some of the difficulties have become systemic in Europe and the United States, in particular due to the slowing down (or even complete halt) of reactor construction, the quantitative reduction in specific training courses, and the slow deindustrialization that is affecting these areas. Many efforts still have to be made in these two areas, based on existing diagnoses and by benefiting from a set of new technologies, relating to the reactors themselves, their calculation, construction and overall organization (links with the States and supply chain included).

By way of example, in SFEN (2018), it has been estimated that the transition from the European prototype stage to a series of EPR reactors (EPR2 concept) could reduce the total overnight investment cost (excluding the interest during construction) by around €4,000/kW in France. Slightly more cautious, RTE (2021) has calculated a value of €4,500/kW. Similar amounts are taken as a reference in the study by OECD/NEA (2020). The latter has adopted the following amounts (see Table 3.3), organized by country and not by type of reactor.

	Technology[8]	Net power (MWe)	Cost ($/kW)
France	LWR (EPR)	1,650	4,013
Japan	ALWR	1,152	3,963
South Korea	ALWR	1,377	2,157
Slovak Republic	Other nuclear	1,004	6,920
United States	LWR	1,100	4,250
Russia	LWR (VVER)	1,122	2,271
India	LWR	950	2,500
China	LWR	950	2,778

Table 3.3. *Overnight costs of future reactors as calculated by the Nuclear Energy Agency (OECD/NEA) in 2019*

In total, the overnight costs of the nuclear facilities to be built are positioned as shown in Table 3.3. On the one hand, we note that cost dispersion is significant (see, for example, the position of Slovakia). They show that the median is around €3,000/kW. There are also disparities between, on the one hand, Europe, the United States and Japan (centered around €4,000/kW), and, on the other hand, China, India, South Korea and Russia (centered around 2,000–2,500 €/kW). This discrepancy is not new. We have already mentioned it in relation to the construction of historic nuclear power. The causes are quite diverse and must be analyzed country by country, even while knowing that the available explanations are not complete. The preponderant factors are the cost of labor, regulatory constraints during construction, the construction pace and serial effects (as well as the industrial environment), monetary parity.

8 LWR: light water reactor, ALWR: *idem* (so-called "advanced" concept).

3.3.2. *Financing terms and risks for parties*

Before concluding on the microeconomic assessment of Generation III nuclear power via the concept of production cost at the station terminal, it is necessary to consider the effects of the rise in uncertainties that currently affect the economy in general (COVID-19, interest rates, questioning of the WTO, etc.) and the production of electricity, in particular. The previous sections recalled that system costs or externalities were apprehensible, but even so, with high uncertainties. The calculation of the LCOE itself does not escape these uncertainties, which were illustrated by the calculation in Chapter 2. It also highlights the major importance of the discount rate. However, this cannot be imposed, but is meticulously chosen depending on the risks affecting reactor financing.

The financing of nuclear power in the future will have a decisive effect on the cost per MWh. This financing is implemented by the project leader, with a generally central role of States, as in the vast majority of large energy investments. The investor raises the necessary capital, either via its own funds or via the markets (through investment banks), or from the States themselves. The conditions for this financing are strongly influenced by the nature of risks. High risks are bearable by the investor only if the project offers a high expected rate of return on investments. It is logical to conduct cost calculations by incorporating a high return on capital, which increases the cost of electricity generated by the reactor. In the event of recourse to the market, the project leader must also face interest rates which (if high) contribute to increasing the cost of capital. To counter these effects, the organization can aim to reallocate the risks to the various stakeholders.

At the collective level, it is socially effective for the States to play a central role in building an electric fleet. This can be done by deciding to bear a more or less significant part of the risks themselves. They do this to ensure the implementation of the conditions for the development of the chosen technologies or, for example, to facilitate an offensive export strategy for their industrial sector. This is typically the case of Russia who has been able to offer significant facilities to countries wishing to equip themselves with VVER technology over the last decade. While they intervene directly in the funding round, States visibly contribute to reducing the overall project risk, and thus facilitate the arrival of other private investors. They can also implement strategies that transfer risks or additional costs to end consumers (as was largely the case for renewables in many countries).

Academic economic research on nuclear financing is relatively underdeveloped. As we have seen, the subject is not easy to tackle due to its specificity, which does not invite numerous original works. While there are several hundred dedicated papers, these are essentially descriptive, with special mention for the financing of long-term waste management (see Chapter 1, Volume 2). Although the compendium by OECD/NEA (2009) offers a broad overview of this literature, it is already old. The report by IAEA (2018) is more recent, but more succinct. Another compendium should be published by the NEA on this subject following the work of the research group set up in 2021.

3.3.2.1. *Attracting investors*

The rate of return required by investors will increase the riskier they perceive the project. A reactor project concentrates various kinds of risks for several billion Euros (high unit amounts), whereas a photovoltaic project worth 15 million euros with support by feed-in tariffs will be low risk. In addition, the risk of building large solar or wind farms can be more easily diversified (locations, technologies, installers, etc.). For many electrical investment projects, the investor seems trapped in a vicious circle: the higher the perceived risk, the higher the required return, the more expensive the investment's financing, the more the projects's total cost increases, together with the market risk. By way of example, Figure 3.7 shows that the cost of a MWh from Hinkley Point in the United Kingdom is multiplied by 2 when the discount rate rises from 3% to 10% (value close to the rate used by EDF for this project).

The financial amounts required for this type of project may require the use of several sources of financing, but may also pose a risk to the balance sheet of the electricity company. As we have seen, one way to attract private investors to the construction of nuclear reactors is to reduce financing costs by reducing project risks, or by transferring all or part of the risks to other players. This issue can be important for States. These can be motivated by the collective interest of a reactor construction project, financed with public rates. But this same project, from the electricity supplier's viewpoint, can be judged negatively and appear unprofitable, because they make their decisions on the basis of much higher private rates. Driven by a logic of collective efficiency, States must set up a policy for sharing such risks, so that the electricity supplier makes a choice based on motivations other than its own.

3.3.2.2. *Nature of the risks for the entities involved in the financing*

Other than the technical risks during construction, other uncertainties weigh on a reactor project, related to the operation of this reactor in the future electrical system. Market risk includes the MWh selling price and its volatility. On the European wholesale market, the average price halved in 10 years (falling from €60 to €30/MWh in the years 2017–2018), then rose notoriously, in the period following the first COVID-19 pandemic, with average values often above €60/MWh, and much higher still at the end of 2021 (over €150/MWh).

This decreasing mechanism was due to the combined effect of falling coal prices, the excess capacity resulting from subsidized investments in renewables and stagnating demand. The ensuing increase can be explained by a shock to gas prices and by the pressure over diminishing production capacities, with a prospect of recovery. This level of volatility is detrimental to investors. In view of mitigating these risks, the principle is to guarantee (in one way or another), a stable valuation level for the amounts of electricity produced. The simplest type of contract is a purchasing power agreement (PPA) at a guaranteed price. This is the case of the Korean contract in the United Arab Emirates.

The political risk associated with the supply of energy is also considerable: a possible challenge by a change of majority, opposition movements delaying construction sites, to mention a few. For nuclear power, it has recently increased in Western countries (including Japan), but remained at a low level in most of Asia and Eastern Europe. However, these risks can be partly mitigated. For the Hinkley Point contract, specific clauses relate to the penalties borne by the British government in the event of a change in the energy policy.

3.3.2.3. *Methods for financing nuclear projects*

The financing modes are described by specifying the players concerned, the legal arrangements, contracts (guarantees), functioning of markets, etc. All of these provisions contribute to the distribution of risks among the players. But the roles of the main actors (States, project leader, operator, other funders, etc.) are diverse, depending on the configuration. For example, a configuration may or may not draw a distinction between the operator and the reactor's owner. The origin of the funds (breakdown between owned and borrowed funds) is also crucial for the investor. The contracts binding these actors are also potentially quite varied. Finally, the actions implemented to secure and stabilize the electricity market – from the perspective of nuclear production – can be of various types. An old historical example which is still pertinent is the regulated prices system in the United States which bills electricity at the cost price, as contracted between the investor and the local authorities.

First, let us remark that the project financing standard model, suitable for many energy projects, has been little implemented in the nuclear sector. This configuration involves establishing a separate company who will own the project; the shares of this company are held by those taking part in the project. This owner company can also obtain loans on its own and afford part of the construction cost, the only guarantee being the shares held by the company itself. In this case, the main purpose is to isolate the risk to diversify it and not to include the project's debt in the funders' accounts. Such a configuration has the advantage of not jeopardizing the shareholders' assets, but is considerably more risky for lenders, whose only guarantee are the assets of the project company itself (ultimately, the reactor). This configuration does not offer enough guarantees for funders (States are generally absent), hence its lack of appeal. As Sainati et al. (2019) have suggested, legal amendments to this model could, however, improve the scenario.

The simplest financing model is the "historical" approach in which nuclear power is managed by the State: it is the nationalized model, notably implemented by France, the United States or the USSR. In this case, the State plays the role of major funder and bears most of the risks: it is the State that is in charge of "absorbing the shocks" of the relative fluctuations in costs and products (it also establishes the market's conditions).

Other models have also been implemented, inspired by project financing with combined funders, but these are based on de facto guarantees provided by the States. A classic pattern is that the State itself (via a public financial institution) takes part in the funding. In such configurations, different levels and forms of public guarantee can be present, so the robustness of the electricity company plays a significant role. The greater this is in terms of the total balance sheet, the more the project's risk is diversified and the lower the financing cost.

A recent trend resulting from the scarcity of financial investors is the contribution of technology vendors, either directly or assisted by their state of origin. Historically, export credit agencies (Coface, US Exim, K-Exim, etc.) had played this role. More recently, it is the companies selling technologies (vendors) that engage banks with public capital, or directly access the capital from electricity companies (in the case of Russia or China).

Thus, Rosatom (Russia) took a 34% capital share in Fennovoima (Finland). Rosatom even went a step further experimenting with the BOOT (Build-Own-Operate-Transfer) model: almost all of the risks (construction, start-up and operation onset) are borne by the seller, which is nonetheless partly backed up via the commitment of the State buyer (who has agreed to a fixed price per kWh guarantee for a contractual period). This modality was chosen in Turkey for the Akkuyu site.

3.3.2.4. *An original example of risk transfer to consumers: the Mankala model*

The Finnish company TVO, which was the first in the world to invest in an EPR, has a special status. It is a private company who sells its electricity on the competitive wholesale market, but has a double originality: its shareholders are mainly large consumers of electricity and it must not generate a profit; it is a cooperative. These shareholders undertake to buy electricity from TVO at the cost price (CMA), either for their own needs (paper manufacturers, chemical industry, etc.) or to resell it on the wholesale market (Fortum, a 25% TVO shareholder), or to resell it to its inhabitants (city of Helsinki, 8% shareholder and other municipalities).

This scheme, called the Mankala model, is not specific to TVO. In 2010, it applied to 42% of Finland's national electricity production. For shareholders, it offers the advantage of a stable long-term MWh purchase price, not subject to price volatility on the wholesale market, and granting supply security. Market risk is low in price and quantity because of the sharing of investment risk among several investor–consumers. Although the resulting cost of capital (WACC) is not public, economic studies published in Finland (Lappeenrata University) on the competitiveness of nuclear power have suggested a 5% discount rate.

3.3.2.5. *Hinkley Point C risk sharing effects on the discount rate*

This project, which concerns two EPRs, was originally part of a high technical risk, considering the other EPR projects in progress and the necessary upgrade of the English nuclear industry, a strong contributor to the project. To counterbalance this effect, it benefited from a low political risk given the consensus among the different political forces on energy policies and a good acceptance of nuclear power among the English public opinion, as well as a limited market risk by the regulations in force. This contemplated the contracts for difference (CfD) applicable to all low-carbon energy projects, guaranteeing an average sale price by compensating for price variations on the wholesale market. This type of approach can be considered a sophisticated form of PPA.

No company with private capital would have taken the risk of such an investment, if the sale price along the facility's operating life had not been protected by a provision of the PPA, CfD or sales contract-type in the long-term, positioning the average selling price at the level of the average discounted cost for the electricity supplier. And even under these conditions the English government justified the high 9% discount rate chosen by EDF energy while determining the guaranteed price for the project to be profitable.

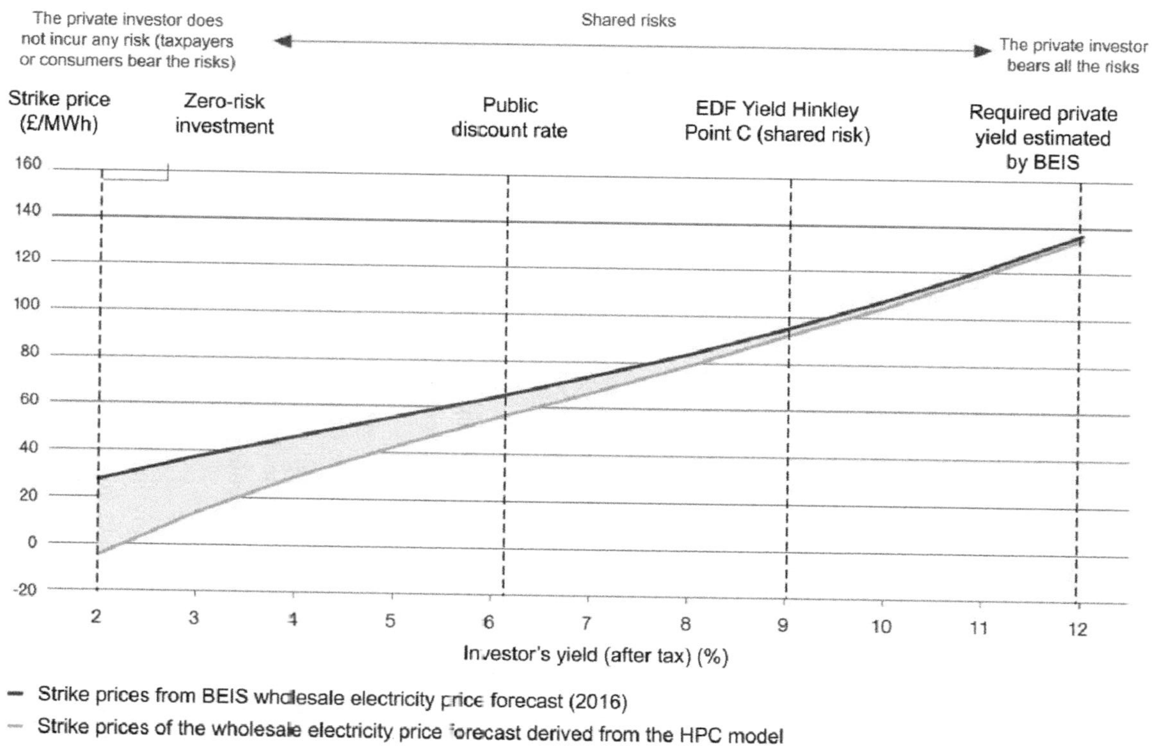

Figure 3.7. *Sensitivity of the guaranteed price according to the investor's rate of return, according to the English Court of Auditors. For a color version of this figure, see www.iste.co.uk/percebois/nuclear1.zip*

Figure 3.7 illustrates the high sensitivity of the strike price (guaranteed price) of Hinkley Point Project A&B projects at the expected rate of return for the investor. Even if the latter is a public investor (merged with the English State), it is currently possible to choose a rate of around 2%, that is to say, a strike price of a few tens of Euros/MWh. With the rate of 9% chosen, the strike price is of around 100 pounds/MWh. This evidences the enormous importance of the contractual architecture and the nature of the risk borne by the stakeholders. Given the strategic nature of these projects, it is socially desirable for them to be implemented. This requires a State guarantee and fairly modest tariffs, or (symmetrically), a tariff sensitivity to the construction cost (left part of the graph). The analysis of the English Court of Auditors shows that, in most scenarios, construction costs could be significantly exceeded before the consumer costs challenge the current agreement, with an ensuing increase in the consumer costs. Assuming that the government needed a (nominal) 6% return, the Court calculated that the construction costs could exceed the initial estimates, shifting from 75% to 100%.

3.3.2.6. *Trends for the future and consequences on the development of nuclear power*

Currently, the rates of return expected by manufacturers in Europe and the United States, taking into account the markets, costs, risks, regulations and related governance are still significantly more important for nuclear power than for renewable energies.

Reducing technical risks (costs, deadlines, etc.) is largely in the hands of manufacturers. This has been partly discussed in the previous sections.

In relation to the other risks, it is important to politically decide on the support that will be given nuclear power, so as to better guide private choices. In particular, this is the challenge posed by the recent discussions on the European “taxonomy” (which public support for which energies?) in order to decide whether or not nuclear power can benefit from the implementation of massively deployed provisions for renewable energies.

States can also act on financing methods (cost of capital) via different means. Take for example the public credit guarantees agreed on the basis of an allowance consistent with the risk borne. Other avenues are also conceivable, such as the creation of “contingent” lines of credit to deal with any changes in safety rules during construction. Another possibility for Europe could be to summon direct loans from the European Investment Bank (EIB) by incorporating subordination elements allocated during the construction period, intended to be refinanced later.

Financial innovation can also take other forms, for example, making the long delays of nuclear projects less penalizing. The English Regulated Asset Based (RAB) system was developed in the United Kingdom (in particular for the Sizewell C project) to finance infrastructure in cases where the first returns on investment appear long after the first disbursements, that is to say, typically over 5 years later. The underlying idea is for investors to be paid since the Construction start date, in anticipation of the electricity that will be delivered in time and then paid according to a tariff set by the regulator.

Finally, another subject also is related to financing as a competition factor. Sellers in OECD countries are regulated in their competition, bound by specific agreements. This is not the case for the aggressive new entrants, such as Russia and China. This provides these countries with the opportunity to offer more advantageous financing and risk-sharing schemes than OECD countries. Traditional sellers and stakeholders of nuclear projects have entered a new field of more complex multilateral competition which requires new strategies for integrating the respective strengths of the protagonists. The international nuclear market is increasingly intertwined. Overlapping competitive systems now coexist, encouraging alliances and partnerships across national borders in order to win nuclear tenders. Funding is a major component at stake in these competitions.

To conclude, contractual configurations may have different forms, but in all cases, they are based on the States (either where the reactors are built, or where the seller is a national actor, see above). This question is broadly comparable to the renewable energy market. Currently, in Europe, almost no electrical investment is made without state support. All of this reinforces the need for a clarification and harmonization of rules, both in relation to the role of States in terms of nuclear power, but also between nuclear power and other low-carbon energies. Concomitantly with the current situation of low real interest rates, this harmonization should also contribute to lowering the costs of nuclear financing in future years.

3.3.3. *Changes in other components of the future nuclear cost*

New reactors generally reduce the costs of other components, as well as those of the reactor itself. These are briefly reviewed below.

3.3.3.1. *Operation and maintenance*

Three key effects reduce operating costs in Generation III reactors. On the one hand, control technologies and operations assistance can make it possible to play on the number of people involved in the various operations (control, maintenance, subcontracting). On the other hand, for certain reactors such as the VVER 1200 and in particular, the EPR, size makes it possible to relate the fixed part of the operating costs to a higher energy produced. Thus, the operating team of an EPR2 unit engages around 400 jobs on site, the order of magnitude of current reactors, which are significantly less powerful.

Finally, maintenance is facilitated by advanced predictive maintenance techniques, in particular those based on the formalization of "digital twins". These simulate the evolution of reactors in real time and make it possible to monitor the condition of physical systems as closely as possible, for better control and adapted maintenance. These techniques also affect the reactor's availability.

At another level, we noticed the recent French difficulties (now under control) and the significant progress in the United States in terms of reducing operating costs. These have borne fruit in recent years, as the report by NEI (2020) has shown. All of this know-how contributes to reducing the operating costs of new reactors and will continue to do so. The bottom-line continues to be the evolution of real wages. This is difficult to assess in the post-COVID-19 crisis phase.

Table 3.4 describes the operating costs used by OECD/IEA+NEA (2020). There is a fairly high variability of a very different nature from that of reactor construction costs. The highest costs tend to be found in Asia (with values around €20/MWh), and the lowest in Russia, in Slovakia, in the United States and in France, to a lesser degree (the levels are around €10/MWh, or a little more for France).

	Operation and maintenance cost	Fuel cost ($/MWh)
France	14.3	9.3
Japan	25.8	13.9
South Korea	18.4	9.3
Slovak Republic	9.7	9.3
United States	11.6	9.3
Russia	10.2	5.0
India	23.8	9.3
China	26.4	10

Table 3.4. *Operating, maintenance and fuel costs for new reactors (NEA 2020; OECD/IEA 2020)*

Compared to the values reported above, describing the costs currently recorded for Generation II nuclear power, these costs show a decrease in Europe, remain stable in the United States and Japan – a country which "traditionally" remains the one with the highest operating costs – and show a relative increase in China. This last phenomenon, perhaps overestimated here, corresponds to the continuous catch-up of Chinese wages compared to major industrialized countries. Likewise, security requirements are increasing. This increase is also noticeable in India. Since these changes affect around a quarter of the discounted production cost (LCOE), these differences partly contribute to rebalancing the differential between Asia and the rest of the world.

3.3.3.2. *Cycle costs (including uranium and back-end, recycling costs)*

The new generation of reactors generally allows a better use of materials (via a higher combustion rate). Because of a slightly higher water temperature, the overall thermodynamic efficiency can also be improved by a few points, thus reducing the relative weight of the nuclear island costs (as well as the relative costs of fuels). In addition, depending on the concepts and the technical enhancements – such as neutron reflectors in the reactor vessel or improvements in the steam generator yields – losses can be prevented, while increasing the electrical power recovered per unit of fuel consumed. Delpech et al. (2004) reviewed the advantages of EPRs in cycle terms. It should also be noted that this reactor would be capable of loading 40% MOX fuel, and that this design could be used for creating 100% MOX cores, in order to diversify recycling strategies.

Prospective studies on the Generation III reactor cycle make it possible to infer cost stability. The only market that has undergone significant fluctuations in the past is that of uranium. In the cycle's front-end, the supply of uranium represents approximately half of the cost. As per the spot price at the beginning of 2021, this cost was less than €2/MWh, because during the last years the price of uranium has fallen on the world market (with a spot price of around \$30/lbU_3O_8). However, this price has recently recovered. The lack of investment in mining resulting from this first situation is starting to show, and the price of uranium is likely to maintain a level of up to \$60/lbU3O_8 for a long time. But the impact on the total production cost will remain low: additional 2–3 €/MWh. In the other front-end stages, it should be noted in recent years that the enrichment technology has shifted from gaseous diffusion, which consumes a lot of energy, to centrifugation, which is much more economical in terms of electricity consumption (factor in the order of 50). This global shift in technology is also lowering the price of the separation unit, which is beginning to translate into the current cycle's costs for decades to come. The other cost components are historically stable. Depending on the case, the impact of technical progress can lead to cost reductions (example of the increase in combustion rates) or moderate increases. The same applies for the new so-called "accident tolerant" tubes, which have a high melting point.

The challenges of the cycle's back-end are hardly technological in the medium term: recycling techniques have been operational for several decades and the design of waste packages is ensured. Costs are known and controlled. For Generation III reactors, as for the historic fleet, a significant issue involves the long-term management of the most active waste. We have already described, as revealed by the Court of Auditors (2014), that the discounted cost for this component is low. As for France, Devezeaux and Berthélémy (2020) have argued that the current cycle will enable a stable cost for at least two decades. Beyond that, another choice in terms of the cycle's back-end would have little long-term effect on the cycle cost.

In the short and medium terms, however, the aforementioned paper has shown that in France, a switch to the open cycle would imply significant additional costs. No decision in this direction is expected. In broad terms, the back-end strategies of the various countries will not notably modify the economics of future nuclear power. In addition, the cycle cost appears to be stable in the future for water reactors, either for the closed cycle or the open cycle.

Table 3.4 describes the values adopted by the OECD. One may appreciate that they are really close for the various countries, which is the characteristic of a globalized market in terms of cycle components. Only Russia appears not to follow this pattern. Compared to the book fuel costs reported for the existing fleet and evaluated within a range of €5 to €8_{2020}/MWh, these amounts may appear slightly higher. This discrepancy is only apparent, and it is in fact due to the discounting effects. In fact, the costs of the services amounting to the fuel purchased by the electricity suppliers are paid a few years before production (cash out). The accounting cycle costs of Generation III reactors will be of the same order as those of the current reactors, if no surprises arise. They should even become lower, as the rate (or scale) effect on fuel production develops (this stage is distinct and, currently, Generation III fuel is produced in still modest quantities).

3.3.3.3. *Other reactor costs*

The other costs, grouped together here, are mainly related to decommissioning. We have previously shown that these costs represent approximately 10–15% of the construction overnight cost. When calculating the discounted cost over the reactor's lifetime, in Chapter 2 we showed that these costs only contribute a few percent to the production cost, due to their late occurrence date. This long cycle makes it possible to accumulate the funds, which will enable decommissioning when the time comes.

The physical nature of Generation III reactors is the same as for the previous generation. The rations should be similar. In addition, the design of new reactors takes into account future decommissioning from the outset and seeks to facilitate it. The costs will be affected with a decreasing tendency. Finally, the costs related to taxes and duties are difficult to anticipate, but their evolution should follow the same trends as for other electrical energies.

3.3.3.4. *Availability and load factors*

Thanks to their endurance and reliability, the new reactor models are designed for high production coefficients, with values of 90% or over, which can help amortize the fixed operating costs. For example, the interruption times for unloading and reloading these reactors are reduced. In the short and medium term, these increased performances will have a direct effect on the production cost, as illustrated in Figure 3.8. In the longer term, if the rate of variable renewables increases notoriously (e.g. over 30, 40% and even further), the preference given to this production means will affect and "chop up" nuclear production, something which is quite feasible because this sector is very flexible. The use rate (production coefficient) during availability will decrease, potentially by several tens of percent for high renewable energy rates. The cost of nuclear power per MWh will increase accordingly.

But, to this inescapable effect, will be added the value of the service rendered by the controllability of nuclear power, which makes it possible to accommodate renewable energies. At present, the remuneration for these services is only partial and relatively modest (via the power guarantee remuneration available through the capacity markets). This will gradually change, both regarding the terms and the price. It will be necessary to remunerate these services at their cost level, under penalty of no longer being able to control the electrical system. The currently cheapest "low carbon" controllable means are gravity hydraulics and nuclear power (both in limited quantities). Beyond that, power-to-gas-to-power batteries or systems are remarkably more expensive. All things considered, revenues from reactors should not decrease, quite the opposite. These developments will be highly context dependent and are still in their infancy. They should provide significant remuneration to the controllable production means in a similar way to the remuneration they will bring to the services provided by electricity storage.

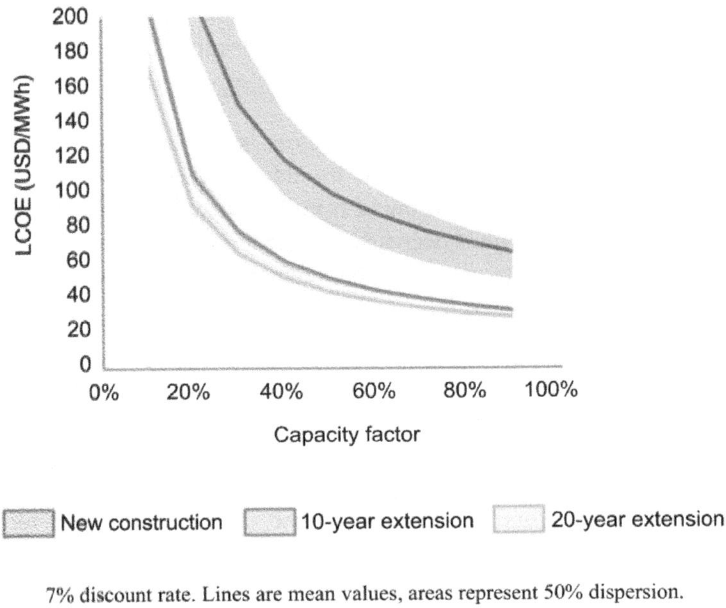

Figure 3.8. *Sensitivity of the discounted cost according to the capacity factor for existing reactors and for Generation III reactors to be built (with a discount rate around 7%) (OECD/IEA+NEA 2020). For a color version of this figure, see www.iste.co.uk/percebois/nuclear1.zip*

3.4. Overview of production costs of other nuclear technologies: SMR and FNR

At the beginning of 21st century, the creativity in terms of new reactor concepts has considerably grown. Not only does this reflect a "rebound" in engineering at the end of the major industrial cycle of the (extensively described) Generation II water reactors, but also a new organization of manufacturers (with the emergence of many start-ups), the access to intensive computing at greatly reduced costs, new possibilities offered by innovative technologies and materials, the availability of large amounts of liquidities supplying venture capital, the arrival of players from countries still not present in this sector, etc. Describing the economic aspects of this wide range of new concepts, while work in this area is still largely at a preliminary draft stage, is not possible here. For the same reason, the topic of nuclear fusion will not be addressed here. The following lines are restricted to families of concepts that are nonetheless beginning to prove mature: certain SMRs and some FNRs.

3.4.1. *SMR reactors*

SMRs have been talked about a lot since the 2010s and there is already a wealth of literature on the subject (OECD/NEA 2021a). Typically, the power of these machines ranges between a few MW and 300 MWe. At the global level, many countries have already demonstrated their interest in SMRs, notably the United States, where the Department of Energy (DOE) actively supports the development and certification of the Musicale reactor. In November 2018, Canada issued a blueprint to mobilize its nuclear industry. France has been working on its own reactor for several years. It has organized itself into a consortium (Nuward) around Technicatome, Naval Group and EDF and is developing a PWR concept of 170 MWe with plate heat exchanger. South Korea has its own project called SMART. It is developing it in cooperation with Saudi Arabia (via KA CARE). Other countries have taken steps to evaluate the interest of an SMR sector or to develop such technology, namely the United Kingdom and Indonesia. The first SMR prototypes of this "new wave" are under construction in Russia and in Argentina (generating power below 50 MWe). The two 35 MWe reactors (RITM-200 type) on the Russian barge Akademik Lomonossov were commissioned in 2020. The first American plant (NuScale, with a 720 MWe capacity, with 12 units of 60 MWe) is under construction in the United States.

Its main technology, ready to be deployed, is PWR. This Generation III technology has a degree of maturity originating in the power reactor and naval propulsion sectors. Submarine or aircraft carrier reactors are typically SMRs, with specific terms of reference. In addition, this technology benefits from an industrial fabric that enables mass production. More broadly, SMR concepts comprise over 50 designs, with a wide variety of design principles and degrees of advancement: practically all historical concepts are present on paper, from salt or molten lead to gas-fired reactors, including single-core units (without refueling).

The interest of SMRs primarily lies in:

– their reduced power level making it possible to broaden the market for nuclear reactors (either by adapting them to modest power demands, or even by offering mobile reactors);

– their modular design, enabling remote manufacturing and faster assembly.

The series effect potentially plays in favor of the economy of these machines. Conversely, their reduced power does not enable a size effect, which has an adverse impact, by increasing the cost per MW. Another disadvantage of these reactors is the lack of maturity of the regulations governing them. Part of the advances in this field still need to be developed, for example, in terms of cross-certifications. Finally, the possible proliferation of these machines in politically unstable geographical areas could create a risk which should be seriously taken into account.

The economy of SMRs is a subject that has given rise to numerous publications, which remain for the most part speculative, because achievements are still in the making. Figure 3.10 generically explains their economic equation. The extrapolation of a concept of large reactors to small sizes (a reduction in size by a factor of 10) would result in an increase in the cost per MW of around 1.5–2.5 for the usual coefficients of the Chilton model presented above. Such a situation would be prohibitive for most of the market. The new interest in SMRs lies in the expected possibility of lowering the costs via basically two approaches. The first approach is related to the concept itself, which should make it possible to benefit from series and pace effects in large factories capable of producing large amounts of prefabricated equipment, to be assembled on site. This type of organization could also reduce construction schedules and the interests during construction.

The second way of operating would be to simplify the concepts (compared to large reactors). A well-known example is that of the melting risks of the PWR tank which go hand in hand with power, making it possible to dispense with a core catcher for medium and small reactors. Other examples include the passive convection cooling mechanisms (without pump) of certain models.

A third approach proposes the extensive standardization of equipment, some of which may even be common to several types of reactors or industrial equipment. Finally, the harmonization of specifications, standards and references within large areas, or even worldwide, could contribute to the desired cost reductions.

All of these changes correspond to a shift in the network of Chilton curves, and the goal is to bring unit costs back to a more acceptable zone. This is schematically described in Figure 3.9.

At present, it is extremely difficult to decide on the economic relevance of such concepts. One of the challenging points is that the market requires a certain size (a number of units of at least 10) to initiate the development of each concept, guaranteeing sufficient amortization of studies and prefabrication plants. As an order of magnitude, we can mention a discounted production cost target of around 100 Euros/MWh for many projects (with a private discount rate). This subject will be clearer in about 10 years.

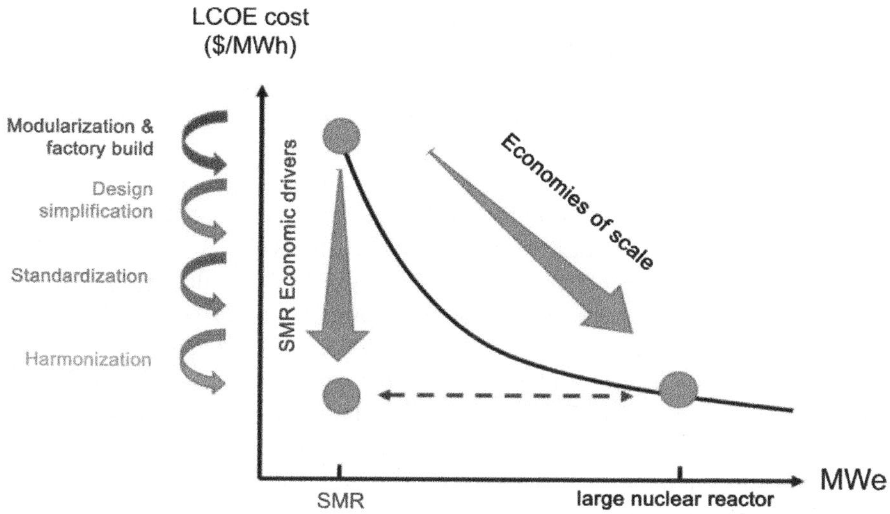

Figure 3.9. *Qualitative presentation of expected cost reduction factors for SMRs in view of coping with to adverse size effects (OECD/NEA 2021a). For a color version of this figure, see www.iste.co.uk/percebois/nuclear1.zip*

The treatment given to SMRs in this book makes it easier to assess their potential market. Both the (unpublished) CEA/I-tésé studies and those of the OECD/NEA show that these reactors should not represent, in the best case, more than a few tens of gigawatts by the middle of this century, while remaining dedicated to specific "niches". The preponderant place we give to large machines is reinforced.

3.4.2. *Fast neutron reactors*

FNRs are available in a fairly large number of technologies and sizes, for example, there are several fast neutron SMR concepts.

These are large reactors (greater than 600 MW). The essential advantage of these technologies is that, via the processing and recycling of fuels and the extraction of the plutonium formed in the reactor, they make it possible to "burn" almost 100% of the natural uranium, whereas slow neutron reactors (essentially water reactors, which slow down the neutrons to facilitate fission) "burn" only around 1% (David 2005). Because of this feature, one can potentially operate fission reactors on a large scale for hundreds of years without the risk of resource depletion. In France, the CEA has extensively worked on these subjects and its economics institute, I-tésé, has produced an abundant literature (CEA 2012, 2015). The main other countries developing these reactors are Japan, Russia and India. The global community working on the Generation IV concepts, which include many FNRs, is grouped under the banner of the Gen IV International Forum (GIF), which also publishes economic assessments on these reactors. For lack of space, this chapter will solely focus on the French case.

At this stage, Generation IV industrial FNRs do not yet exist, and it will take at least 20 years for the first ones to be authorized by the safety authorities and built. The results of the current estimates of the CEA and EDF are that the construction cost of these reactors, for a size of 1,000 to 1,500 MW, could be 30% higher than that of water reactors. That is around 5,000–6,000 €/kW. These assessments relate to series reactors and will have to be refined as the design work progresses. The fuel cycle is less expensive though, mainly because the consumption of fissile material will use up available stocks. As shown by Devezeaux et al. (2012), there is no need to obtain natural uranium to supply these reactors, because the current stocks of depleted uranium from the water reactor's front-end cycle will suffice for centuries. In total, given the current costs, the LCOE of a MWh generated by a fast reactor appears to be at least 20% more expensive than that of Generation III reactors. But one should not deduce that these orders of magnitude definitively invalidate this family of technologies, compared to the Generation III water reactors. In the long-term (i.e. 50 years and over), the following points must be taken into account:

– The logic is not necessarily to compare a fleet entirely made up of FNRs with a fleet of water reactors, but with so-called "symbiotic" fleets that combine these two types of reactors and make it possible to completely cease uranium consumption for a minimal investment in reactors. The presence of FNRs is only minor to such an extent that the fleet's total cost (cost of the reactors alone) is not much higher, whereas the cycle cost can cancel out the uranium component.

– FNRs have advantages in terms of fluid temperature and neutron flux which enable them to provide services inaccessible to water reactors (process heat supply, in particular).

– The "reoptimization" of FNRs offers greater potential than water reactors due to their lesser advancement in the innovation cycle and the technical lifecycle. This is currently in progress.

One should also bear in mind that significant uncertainties affect current valuations, if only because the economic interest of these machines will not materialize, if ever, before the second half of the century, as Baschwitz et al. (2017) have pointed out. With this horizon in mind, the energy landscape will have remarkably changed, both in terms of reactor applications and competitive performance, as well as in relation to the new *optima* to be attained within electrical and energy systems not to mention in terms of the nuclear technologies themselves (new materials, modularity, calculation, etc.). In short, the emergence of these reactors will primarily depend on the capital costs and the increase rate in the fuel cost of water reactors, following the probable increase in the price of uranium in the post-2050 horizon. At the same time, the other determinants of the cost of nuclear power will play their role, as has been analyzed for historic nuclear power. Technical progress, changes in the services provided, safety requirements and the cost of labor will influence the costs of the nuclear fleets and the other components of the electrical fleets.

All other things being equal, and considering that the price of uranium should increase from around $400 to $500/lb to sparkle economic interest in a mixed fleet of water and of FNR reactors, the production cost of the nuclear fleet could rise by around 10% (CEA 2015). Such a gap, projected by the second half of the century, does not seem to represent much in view of all the other uncertainties on the determinants of costs (and a fortiori, the choices to be made). The long-term competitiveness of fast reactors, within a mixed fleet, is today impossible to anticipate appropriately. As a result, the current research and development in these technologies are carried out following an insurance logic by countries that have a strong long-term interest, such as China, which has decided to build its own fuel recycling plant and to develop Chinese FNR technology.

In summary, while Generation IV FNRs offer promising features in the medium and long-term, their development will probably be minor until the end of this century. Their competitiveness is still poorly known due to a maturity still to be built within the framework of the generation to come. On a similar note, there is still a great deal of uncertainty regarding the physiognomy of the future electric fleets that will accommodate these machines, and the market rules that will govern them. The CEA (2015) has assessed their potential market at a few dozen units by the mid-century, in the context of large fleets or possible international "hubs" specializing in fast reactors and the related plutonium cycle. The preponderant place given in this work to water reactors and large-scale slow neutron reactors (known as "thermal" units) is confirmed by this analysis.

3.5. Electrical system costs and nuclear competitiveness

The costs described above, and summarized via the LCOE, are future costs considered by a given economic agent under certain conditions. At the beginning of this chapter, we explained that the analysis by accounting costs or LCOE, which was the rule "during the previous century", is no longer sufficient today. There are two main reasons for completing the analysis:

– The existence of significant system costs within electrical systems.

– The existence of externalities which must be integrated into the calculations as much as possible. A recent example of the explicit quantification of system costs has been provided by the RTE (2021) study, within the framework of an analysis of electricity scenarios for France.

This topic will only be addressed succinctly, focusing on the tools for taking them into account operationally, the quantitative results obtained and the way in which these results profoundly modify the assessment of the structural choices for future fleets.

No source of electricity generation is perfect. Each being limited to (a) their geographical production capacity (rare are those which are exactly positioned at the place of the demand), (b) their availability and short-term predictability (the source is affected by hazards and its flexibility is limited), or (c) medium-term availability (its deliverability is never 100%, if only for maintenance). The identification of these system costs (Joskow 2011; OECD/NEA 2012a; Hirth et al. 2015) aims to assess the costs brought about by these constraints and historically goes back to acknowledging the variability of renewable energies, which no longer made it possible to compare the variable production must run and controllable production without additional precautions, when significant shares of the electricity originated in such production means. In total, the nomenclature of these costs can be summarized into three main components:

– The costs of connecting production and network units (lines), which may vary depending on the spatial distribution of the production and the demand (particularly during the peak).

– The costs related to short-term supply-demand balancing, which is modified by the existence of the studied equipment subject to short-term forecast errors on deliverability and is more or less capable of contributing to this equilibrium (in the face of stochastic demand), particularly in terms of flexibility.

– So-called “profile” costs, resulting from the addition of a new unit to the existing fleet. They are related to the mid-term or long-term temporal profile of production units and their hourly availability (example of solar PV generators and day/night alternation, or nuclear energy during fuel reloading).

Apart from the connection costs, the concept of system costs associated with an equipment is only correctly defined for optimal fleets at equilibrium (minimum cost). In practice, this notion is often extended by comparing “rigid” (and non-optimal) fleets comprising more or less renewables. These system costs are not zero for controllable means, and they apply to any means of energy production or storage. But they are (much) more important for variable energies and require being taken into account for serious comparisons. Unfortunately, these system costs are difficult to determine: not only do they depend on the geography (wind, sunshine, network, etc.) and the country and/or network, but also on the fleet's structure and share of nuclear energy (essentially controllable energies) and variable energies in the electricity mix, as well as on the electricity demand. Note also the crucial role of hydraulics, with its various forms, in a fleet like the one found in France.

3.5.1. *Calculation of system costs*

The calculation of connection costs is specific to each facility. The calculation of other costs requires the provision of a fleet model, such as the Antares suite used by RTE. Moreover, these costs strongly depend on the fleet's structure, which is never optimal. For example, the value (the inverse of the cost) of a controllable type of energy such as nuclear is all the higher, the higher the share of non-controllable energy (solar, wind). In addition, these costs depend on a host of fairly numerous and complex data and hypotheses (for the future), such as the condition of connected systems in Europe or meteorological and rainfall statistics. A final subtlety for this type of analysis is that the formal calculations are generally carried out for fixed fleets (static analysis). One way to address this problem is to provide cost values every 5 or 10 years, as the fleet's structure evolves.

Moreover, the existence of such costs raises the question of their translation into the price system. These costs are not externalities and therefore affect electrical systems: when they increase, they increase the cost per MWh. But there are few mechanisms available to correctly assign them to the generating event. The economic logic would be to allocate such costs to each production means once they have been measured, each to the extent of its responsibility. Similarly, each piece of equipment reducing them should be remunerated for the services rendered. This is, in very broad terms, the way it happens with all electrical entities anywhere in the world. On the other hand, the public calculation must acknowledge them when choosing the fleet and impute them to the productions generating them.

3.5.2. *Nuclear system costs*

Nuclear system costs are low. A first part of this positive result is due to the connection costs which are generally well controlled by positioning reactors near high voltage lines. If this is not possible (e.g. at Flamanville), the costs may be higher. A second part of this result is due to the mechanical inertia of the alternators of nuclear power plants, which strongly contributes to stabilizing the network in the short-term. A third part is due to the flexibility of nuclear reactors, which is often overlooked. The OECD/NEA (2012b) devoted an entire report to tithe performances of nuclear power are comparable to those of fossil fuel power plants, such as combined cycle gas power plants. The power gradient can reach 5% per minute. This is what is currently making an important contribution to supporting the rise in power of non-controllable variable energies.

3.5.3. *System costs and choice hierarchy*

The progress made during the last decade is major and provides orders of magnitude complementing and significantly modifying the results of LCOE calculations. The main lessons learned from this work are summarized as follows:

– The costs of the nuclear system and controllable equipment are low (around €2–3/MWh according to the OECD/NEA (2012a) for a French-type fleet), and only slightly depend on the share of this equipment for total power or energy.

– System costs for variable energies (essentially solar PV and wind power) are much higher, depending on the case, and can reach several tens of euros/MWh (typically up to €50/MWh for a French- type facility with 75% variable renewables).

– These costs strongly depend on the share of this equipment for total power or energy, and increase with it. They will decrease over time with the arrival of innovations and the downward trend in short-term storage costs.

– In the coming decades, it is probable for the costs of variable renewables to continue to fall sharply. This trend will enable their continued development and mitigate the increase in system costs. These opposing forces could converge around 2050 toward renewable energy rates (hydraulics not included), which can be described as “optimal”, located around 30–40% and even beyond (up to 80% in RTE’s (2021) M1 scenario), the rest being made up of controllable sources, including nuclear and gravity hydraulics.

The literature now abounds in calculations of this type. A summary can be found in Samadi (2017) or in OECD/NEA (2018, 2019). Table 3.5 shows a compilation of numerous data from the literature providing orders of magnitude, without the extreme values. The table illustrates the system costs in networks, integrating important flexibility possibilities (technology, exports, etc.). The costs of the technologies underlying the studies this summary is based on are essentially prospective costs, the only ones that can be envisaged for large shares of nuclear or renewable energy.

The main result is that nuclear energy has always had low system costs (as the other controllable sources not described here). Variable renewables have system costs generally increasing with their contribution to production, but with levels which can remain relatively moderate in some cases, even for high renewables rates (around twenty euros/MWh). However, these costs can turn out to be much higher in other cases (over €50/MWh). These orders of magnitude show that the approaches at the power station terminal or the *grid parity* type of reasoning are obsolete. One of the major challenges of R&D is, and will be, to make these system costs decrease.

	Shares in production			
	10 %	30 %	50 %	70 %
	Nuclear			
Connection + Grid	0–2	0–2	0–2	0–2
Balancing + profile	0	0	0	Not significant
	Wind power			
Connection + Grid	3–10	3–20	5–25	Not significant
Balancing + profile	4–10	7–20	10–25	Not significant
	Solar power			
Connection + Grid	3–10	3–15	3–20	Not significant
Balancing + profile	5–15	10–30	10–35	Not significant

Table 3.5. *Summary of orders of magnitude (in €/MWh) of nuclear, wind and solar system costs. These data are general, as the situations differ depending on the technology, the location and the fleets (OECD-NEA 2012a; Strbac et al. 2016; Samadi 2017; OECD-NEA 2018, 2019, 2020)*

Another important result from the nonlinearity of the total costs according to the fleet's structure is that the optimum (if there is an optimum and if it is possible to reach it) includes both variable renewables, hydraulics, controllable low-carbon resources (typically nuclear power or equipment burning fossil fuels with carbon capture and storage). There is no mutual exclusion between these production means and their management, but a form of complementarity. Note that "100% Renewable Energy" scenarios such as those described by Ademe (2015) achieve such extreme scores because of bold hypotheses that often avoid the prohibitive costs of extreme or "corner case" scenarios. For us, they tend to look more like conceptual exercises rather than cases to be taken into account in the perspective of concrete planning.

3.5.4. *Competition, synergies and complementarities between nuclear and renewable energy*

At first, the combination of nuclear and renewable low-carbon energies was not conceived as such. As has been shown, the effects of systems within fleets with a high proportion of variable renewables lead to cuts in the production of nuclear reactors. These are no longer summoned part of the time, because wind turbines and solar panels supply the entire demand at zero variable cost. These phenomena, inherent in the temporal distribution of production and its variability, cause a decrease in the production coefficient (utilization rate) of nuclear power, and an increase in its production cost. Conversely, the services rendered by nuclear power during hours of high residual demand (deducted from renewable energy) have a higher value and nuclear power must be remunerated accordingly. While the increase in renewable energy levels is rapid – within a context of weak growth (or decrease) in the average demand – the fleet cannot readjust, and the loss of energy from nuclear power becomes a "deadweight loss" for the operator.

But, as shown in Figure 3.10, in the long-term dynamic, there appears to be a form of convergence between these energies. The average values of the (production + system) costs from nuclear power and renewable energies (wind and solar at parity) for a country like France (with 20% hydraulics) are likely to converge by 2050. One may appreciate that the energy cost curves intersect in an area around 15–45% for renewables. This figure shows that approaching the future fleets via the LCOE is not sufficient because system costs are contingent and vary greatly in function of penetration rates. The curvature on the right is also sensitive to factors such as the overall decarbonization constraint, whether the electrical network studied is open (or not), or the limits of the hydraulic or renewable energy deposits in function of the annual energy. The major conclusion to be drawn from these approaches is that the minimum cost is achieved via a mix including renewables and nuclear. It corroborates the several forms of complementarity between nuclear power and variable renewables (Devezeaux and Brière 2017).

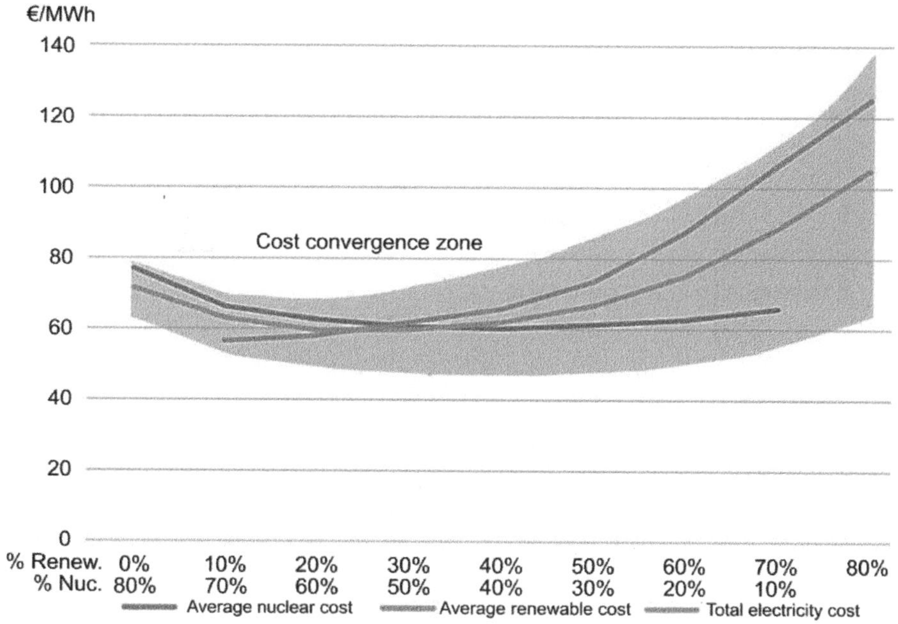

Figure 3.10. *Order of magnitude of production costs, nuclear and renewable energy systems in a French-type electricity mix by 2050. Author's calculations based on previous results and data (OECD/NEA 2019). For a color version of this figure, see www.iste.co.uk/percebois/nuclear1.zip*

3.6. Conclusion

History has shown that nuclear costs are not a "long quiet river" (see Figure 3.11).

In its early days, nuclear power went through an intense phase, burgeoning with diversified concepts, which underwent an acute "internal Darwinism" which resulted in powerful cost reductions, due to the rise in size and early learning effects.

Then followed the "historic" Generation II nuclear cycle, currently implemented in Europe and the United States. This paved the way for a new convergence and strong standardization: water reactors have emerged as winners (mainly pressurized reactors). Generation III standards were the result from this cycle, framing machines which are not too diverse, as far as high-power reactors are concerned.

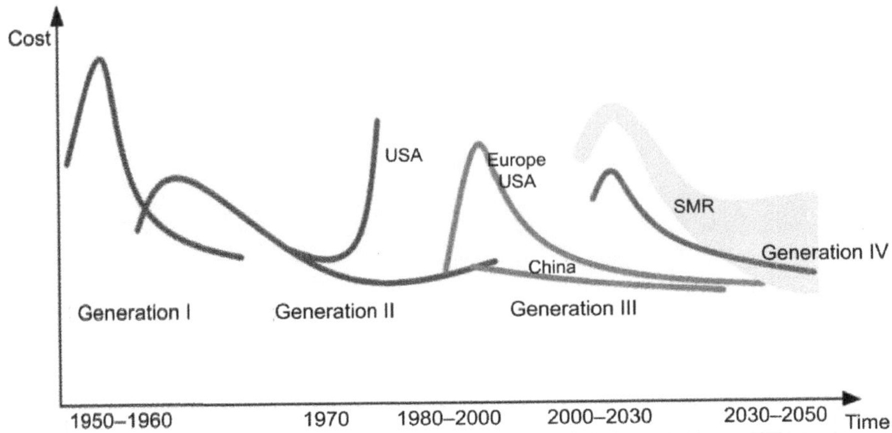

Figure 3.11. *Stylized succession of nuclear generations and profiles of their overnight investment costs per kWe*[9] *(source: author's compilations and calculations). For a color version of this figure, see www.iste.co.uk/percebois/nuclear1.zip*

In concrete terms, this historic Generation II amounts to the vast majority of the current fleet. Its FEC is approximately €60/MWh in France, which confirms, after several decades, the choice of this technology against other alternatives. Even if this evaluation can be used to set a selling price for historic nuclear electricity (ARENH issue), other indicators are operationally even more important. The annual average cash costs have remained low (around €30–40/MWh in France) and ensure the benefit of operating the fleet almost everywhere. They are much lower than those of any controllable technology that could replace historic nuclear power in OECD countries. Moreover, the economic studies on the extension of the productive life of reactors show the almost systematic interest of making the related investments, rather than shutting down the reactors. According to the OECD/NEA, the LCOEs of this energy are around 25–40 €/MWh for a 20-year extension. Finally, the short-term marginal costs are extremely low (around €10/MWh), which ensures a regular call for production from existing reactors.

Currently, the technology of large Generation III reactors has completed its convergence phase. The articulation between Generations II and III dates back to the 1990s, when continuity prevailed, particularly in terms of standards and methods.

9 Generation IV is that of sodium FNRs alone. We should also note that the evolution in the costs of SMRs remains uncertain.

On the other hand, the sluggishness of the orders during two decades has not boosted the design capacity of OECD countries. In parallel with this evolution, not to say concentration, quite linear and with a long time constant, the world has rapidly changed and offered new opportunities (technological, political, industrial, etc.). These are now fully intertwined with nuclear power. We are witnessing powerful transformations, which either remain within the technical and regulatory framework of Generation III (with altogether largely defined concepts), or directly result from them, with SMRs and the Generation IV concepts.

In relation to these new concepts, in certain aspects, the situation turns out to be similar to the beginnings of nuclear power, in the sense that a potential revolution is preparing and the number of technical options (re)opened is abundant. It is difficult to estimate the concrete translation (in number of units) of all these technologies and the associated revisions of the reactors' construction methods. What is certain is that this creativity, combined with research opening up to new fields, brings and will bring new ideas and solutions, some of which could also be useful for large reactors. In comparison with solar power, or wind power, the nuclear Generation III is in a situation of fairly strong technical uniqueness, so that only incremental progress is possible for the moment. The coming period will potentially enable nuclear power to make further radical progress. But the characteristic times for nuclear power development are still long and the quantitative multiplication of SMRs, if it ever succeeds, will not take place before the 2040s. It is also with this horizon in mind that new cost reductions, resulting from these breakthrough approaches, will be able to spread throughout the nuclear industry, even to large reactors.

Finally, changes in regulations, international standards and mutual certifications (as in aeronautics) could make it possible to reduce the distinctive long time frames, which are often too penalizing to achieve efficiency, as we have seen in the field of construction.

For the short- and medium-term future, the main determinants of the costs per MWh at the power station terminal have already been presented. They seem to announce a stable trend in these costs or even a decrease. The role of cycles, of various kinds, will not disappear however. Apart from an increase in commodity prices in the ongoing global recovery movement after the strongest 2 years of the Covid-19 epidemic, these cycles appear rather favorable to cost moderation.

In very concrete terms, we have amply illustrated the major difficulties for Generation III in Europe. The nuclear industry in OECD countries is compelled to demonstrate the reduction in construction costs in the years to come. For this, we have seen the importance of enrolling constructions in a program of a certain size (e.g. the six EPR2 in France decided at the end of 2021) and, apart from that, the key role of learning effects. This is the great challenge of the coming decade. Otherwise, Asia will become the global leader in this industrial sector.

In this view, the new water-powered Generation III power plant appears competitive, in terms of LCOE, compared to other sources of electricity production, in most regions of the world. According to the OECD/NEA, its LCOE is around €40–70/MWh. This assessment is extensively documented and comprehensive. For the electricity supplier in France, the LCOE of an EPR2 type of reactor should be positioned in the €60–70/MWh range (with a discount rate of up to 7%).

Finally, note that the future of nuclear energy in the energy transition cannot be exclusively approached with the tools presented in this chapter, and in particular, the LCOE. Figure 2.1 reminds us of this (the analysis is here restricted to its “core”, that is, at the power station’s terminal). To go further, it is imperative to take into account the effects (the costs and benefits) within the electrical systems. This is a real revolution in the economy of electricity, still often poorly known and understood, but with first-rate consequences. We have shown that the literature, still in progress, makes it possible to give reliable orders of magnitude, even if they are not universal.

Figure 3.12 combines the results of LCOE and system costs, for a country close to France (assuming 20% of the energy comes from hydraulics), at a distant horizon (2050). We have plotted the shape of the total cost of the electricity fleet’s production, including the system costs, which shows a strong concavity upwards. It follows that there is an optimal zone (where the costs are the lowest), which is positioned for nuclear power located in the range of 30–70%, that is, for 15–45% of renewables (here wind and solar equally). This field is wide and will have to be refined according to technical developments and related costs. But what this approach shows is that extreme choices, especially those on the right of the graph, can lead to considerable additional costs. In the case where the costs are the lowest (renewables are deemed to reach levels of €30–35/MWh by 2050), the right part of the graph shows results close to those by RTE (2021), with an additional cost of around 20% for a fleet made up almost exclusively of renewable energy (wind and solar) and hydraulics. The top curve is characterized by little technical progress and strong constraints on greenhouse gas emissions for advanced equipment and the adequacy of supply and demand.

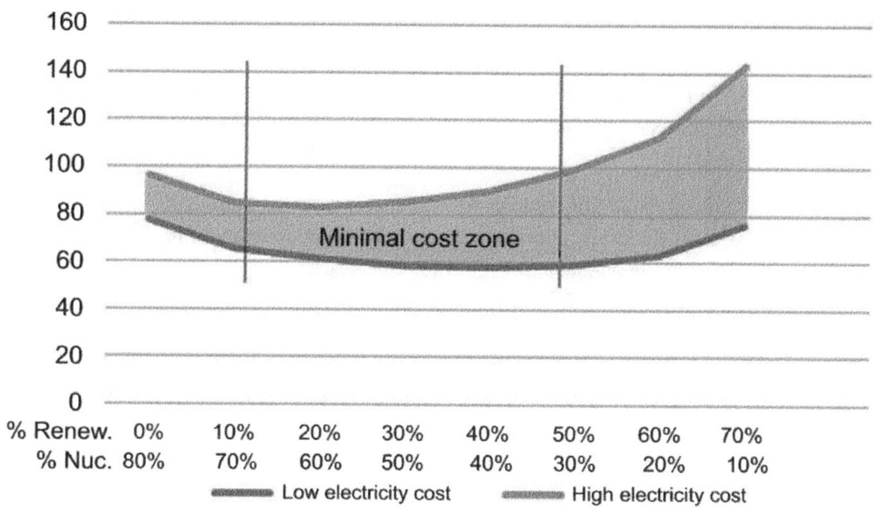

Figure 3.12. *Order of magnitude of the costs of a French-type electrical system by 2050. Author's calculations on the same basis as shown in Figure 3.10. For a color version of this figure, see www.iste.co.uk/percebois/nuclear1.zip*

These results show that a cost comparison with a fixed fleet structure, even if we include system effects, says little about the choices to be made. In any case, the combination of renewables and nuclear is the best solution.

Finally, in addition to system costs, Chapter 2 showed that other types of costs deserve to be taken into account: climate externalities (which are nonetheless negligible if the fleet is carbon free), health impacts of gaseous and liquid effluents, waste, accidents, etc.

These aspects are beyond the scope of this chapter, but the book (OECD/NEA 2018) covers these issues and, via the notion of full social cost, shows that nuclear energy is also well positioned, with one of the lowest costs among all the sources of electricity.

From an economic perspective, the use of nuclear power in many regions of the world, in complementarity with a growing renewable energy base, seems solidly founded. But the development of this energy still depends (and will continue to do so), on the wishes of the States and/or regions and territories. These choices have a direct impact on financing methods and on the allocation of electricity project risks (market risks in particular). They are underway in Europe and will have a decisive impact.

3.7. References

ADEME (2015). Un mix électrique 100% renouvelable ? Analyses et optimisations [Online]. Available at: https://www.ademe.fr/sites/default/files/assets/documents/mix-electrique-rapport-2015.pdf.

Baschwitz, A., Gabriel, S., Mathonnière, G., Devezeaux de Lavergne, J.-G. (2017). When would fast reactors become competitive with light water reactors? Methodology and key parameters. *Progress in Nuclear Energy*, 100, 103–113.

Bertel, E. and Naudet, G. (2004). *L'économie de l'énergie nucléaire*. EDP Sciences. Les Ulis.

Berthélémy, M. and Escobar, L. (2015). Clear reactors' construction costs: The role of lead-time, standardization and technological progress. *Energy Policy*, 82, 118–130.

Boccard, N. (2014). The cost of nuclear electricity: France after Fukushima. *Energy Policy*, 66(C), 450–461.

CEA (2012). Séparation-transmutation des éléments radioactifs à vie longue, Tome 2. Report for the Government, Direction de l'Énergie Nucléaire.

CEA (2015). Avancées des recherches sur la séparation-transmutation et le multirecyclage du plutonium dans les réacteurs à flux de neutrons rapides. Report for the Government, Direction de l'Énergie Nucléaire.

Champsaur, P. (2012). Rapport de la Commission "Champsaur II". Report for the Government.

Cour des comptes (2012). Les coûts de la filière électronucléaire. Report, Cour des comptes, Paris.

Cour des comptes (2014). Le coût de production de l'électricité nucléaire. Actualisation 2014 du rapport de 2012. Report, Cour des comptes, Paris.

Cour des comptes (2016). Rapport public annuel. Report, Cour des comptes, Paris.

Cour des comptes (2020a). L'arrêt et le démantèlement des installations nucléaires. Report, Cour des comptes, Paris.

Cour des comptes (2020b). La Filière EPR. Report, Cour des comptes, Paris.

D'Haeseleer, W.D. (2013). Synthesis on the economics of nuclear energy. Study for the European Commission, DG Energy.

David, S. (2005). Future scenarios for fission-based reactors. *Nuclear Physics A*, 751, 429–441.

Debès, M. (2002). Les avantages industriels et économiques de l'accroissement des taux de combustion : le point de vue d'EDF. *Revue générale nucléaire*, 1, 64–69.

Delpech, M., Carlier, B., Devezeaux de Lavergne, J.-G., Debes, M. (2004). Le réacteur EPR et le cycle du combustible. *Revue générale nucléaire*, 6, 65–69.

Devezeaux de Lavergne, J.-G. and Berthélémy, M. (2020). Coûts du nouveau nucléaire et éclairage sur l'économie du cycle. *Annales des Mines, Responsabilité et environnement, Le nucléaire civil, enjeux et débats*, 97(1), 35–47.

Devezeaux de Lavergne, J.-G. and Brière, E. (2017). Nucléaire et énergies renouvelables : des technologies complémentaires pour la transition énergétique. *Revue générale nucléaire*, 1, 8–13.

Devezeaux de Lavergne, J.-G. and Durret, L.-F. (1995). Économie du plutonium. *Revue générale nucléaire*, 1, 40–45.

Devezeaux de Lavergne, J.-G., Gauché, F., Mathonnière, G. (2012). Recyclage du combustible et réacteurs de quatrième génération. *RGN*, 2, 92–99.

Devezeaux de Lavergne, J.-G., Doan, L., Duquesnoy, T. (2020). Stockage des déchets radioactifs en profondeur : que décider ? Quand décider ? *Revue de l'énergie*, 650, 27–48.

Duquesnoy, T. (2013). Coût de construction des réacteurs REP : évolution des conditions économiques ou accroissement de la complexité ? [Online]. Available at: http://i-tese.cea.fr/_files/LettreItese18/ECLAIRAGES/REP.pdf.

EPRI (2010). Program on technology innovation: Integrated generation technology options. Report, Electric Power Research Institute (EPRI) 3420 Hillview Avenue Palo Alto, CA.

Escobar, L. and Lévêque, F. (2015). Revisiting the cost escalation curse of nuclear power: New lessons from the French experience. *Economics of Energy & Environmental Policy* [Online]. Available at: https://hal-mines-paristech.archives-ouvertes.fr/hal-00780566/file/I3WP_12-ME-08.pdf.

Folz, J.-M. (2019). La construction de l'EPR de Flamanville. Report to the Président Directeur Général of the EDF.

Grubler, A. (2010). The costs of the French nuclear scale-up: A case of negative learning by doing. *Energy Policy*, 38(9), 5174–5188.

Hansen, J.-P. and Percebois, J. (2019). *Énergie : économie et politiques*. De Boeck Supérieur, Louvain-la-Neuve.

Hirth, L., Ueckerdt, F., Edenhofer, O. (2015). Integration costs revisited – An economic framework for wind and solar variability. *Renewable Energy*, 74, 925–939 [Online]. Available at: https://doi.org/10.1016/j.renene.2014.08.065.

Hogue, M. (2012). A review of the costs of nuclear power generation. Working paper, Bureau of Economic and Business Research [Online]. Available at: http://bebr.business.utah.edu/sites/bebr/Documents/studies/Nuclear_Report_Final_Web_7Mar2012.pdf.

IAEA (2018). *Economic Assessment of the Long Term Operation of Nuclear Power Plants: Approaches and Experience*. IAEA Nuclear Energy Series, Vienna.

Joskow, P.L. (2011). Comparing the costs of intermittent and dispatchable electricity generating technologies. *The American Economic Review*, 101(3), 238–241 [Online]. Available at: http://www.jstor.org/stable/29783746.

Joskow, P.L. and Parsons, J.E. (2009). The economic future of nuclear power. *Daedalus*, 138(4), 45–59 [Online]. Available at: http://www.jstor.org/stable/40544000.

Kee, E. (2010). Global nuclear power developments – Asia leads the way. In *Nuclear Energy Asia 2010 Conference*, Hong Kong.

Koomey, J. and Hultman, N.E. (2007). A reactor-level analysis of busbar costs for US nuclear plants. *Energy Policy*, 35(11), 5630–5642.

Koomey, J., Hultman, N.E., Grubler, A. (2017). A reply to "Historical construction costs of global nuclear power reactors". *Energy Policy*, 102, 640–643. 10.1016/j.enpol.2016.03.052.

Leprince, P. and Popiolek, N. (2016). La dépense de R&D en quelques chiffres clé : éclairage particulier sur le secteur énergétique. *Revue de l'énergie*, 632, 309–320.

Locatelli, G. and Mancini, M. (2011). The role of the reactor size for an investment in the nuclear sector: An evaluation of not-financial parameters. *Progress in Nuclear Energy*, 53(2), 212–222.

Locatelli, G. and Mancini, M. (2012). Looking back to see the future: Building nuclear power plants in Europe. *Construction Management and Economics*, 30, 623–637.

Lovering, J., Yip, A., Nordhaus, T. (2016). Historical construction costs of global nuclear power reactors. *Energy Policy*, 91, 371–382.

Lovering, J.R., Nordhaus, T., Yip, A. (2017). Apples and oranges: Comparing nuclear construction costs across nations, time periods, and technologies. *Energy Policy*, 102(C), 650–654.

MIT (2009). The future of nuclear power: An interdisciplinary study. Report, MIT, Cambridge, MA.

MIT (2011). The future of the nuclear fuel cycle: An interdisciplinary study. Report, MIT, Cambridge, MA.

Moynet, G. (1984). Évolution du coût de l'électricité en France au cours des dix dernières années. *Revue générale nucléaire*, 2, 141–153.

NEI (2020). Nuclear costs in context. Report, Washington, D.C.

NIRAB (2020). Achieving net zero: The role of nuclear energy in decarbonisation. Annual report. NIRAB, UK.

OECD/IEA+NEA (2015). Projected costs of generating electricity. Report, IEA (International Energy Agency), Paris/NEA (Organisation for Economic Co-operation and Development/ Nuclear Energy Agency), Boulogne-Billancourt.

OECD/IEA+NEA (2020). Projected costs of generating electricity [Online]. Available at: https://www.iea.org/reports/projected-costs-of-generating-electricity-2020.

OECD/NEA (1994). The economics of the nuclear fuel cycle [Online]. Available at: https://www.oecd-nea.org/upload/docs/application/pdf/2020-07/the_economics_of_the_nuclear_fuel_cycle_1994.pdf.

OECD/NEA (2009). The financing of nuclear power plants [Online]. Available at: https://www.oecd-nea.org/upload/docs/application/pdf/2019-12/financing-plants.pdf.

OECD/NEA (2012a). Nuclear energy and renewables: System effects in low-carbon electricity systems, nuclear energy agency [Online]. Available at: https://www.oecd-nea.org/upload/docs/application/pdf/2019-12/system-effects-exec-sum.pdf.

OECD/NEA (2012b). The role of nuclear energy in a low-carbon energy future, nuclear energy agency [Online]. Available at: https://www.oecd-nea.org/upload/docs/application/pdf/2019-12/nea6887-role-nuclear-low-carbon.pdf.

OECD/NEA (2013). The economics of the back end of the nuclear fuel cycle [Online]. Available at: https://www.oecd-nea.org/upload/docs/application/pdf/2019-12/7061-ebenfc.pdf.

OECD/NEA (2015). Nuclear new build: Insights into financing and project management [Online]. Available at: https://www.oecd-nea.org/upload/docs/application/pdf/2019-12/7195-nn-build-2015.pdf.

OECD/NEA (2018). The full costs of electricity provision [Online]. Available at: https://www.oecd-nea.org/ndd/pubs/2018/7441-full-costs-2018-es.pdf

OECD/NEA (2019). The costs of decarbonisation: System costs with high shares of nuclear and renewables [Online]. Available at: https://www.oecd.org/publications/the-costs-of-decarbonisation-9789264312180-en.htm.

OECD/NEA (2020). Unlocking reductions in the construction costs of nuclear: A practical guide for stakeholders [Online]. Available at: https://www.oecd-nea.org/upload/docs/application/pdf/2020-07/7530-reducing-cost-nuclear-construction.pdf.

OECD/NEA (2021a). Small modular reactors: Challenges and opportunities. [Online]. Available at: https://www.oecd-nea.org/upload/docs/application/pdf/2021-03/7560_smr_report.pdf.

OECD/NEA (2021b). Long-term operation of nuclear power plants and decarbonisation strategies [Online]. Available at: https://www.oecd-nea.org/upload/docs/application/pdf/2021-03/7560_smr_report.pdf.

Parsons, J. and Du, Y. (2003). The future of nuclear power: An MIT interdisciplinary study. MIT technical report, MIT, Cambridge, MA.

Percebois, J. (1989). *Economie de l'énergie*. Economica, Paris.

Percebois, J. and Mandil, C. (eds) (2012). Énergies 2050. In *Rapport au gouvernement*, La Documentation française, Paris.

Poinssot, C., Bourg, S., Ouvrier, N., Combernoux, N., Rostaing, C., Vargas-Gonzalez, M., Bruno, J. (2014). Assessment of the environmental footprint of nuclear energy systems. Comparison between closed and open fuel cycles. *Energy*, 69, 199–211 [Online]. Available at: https://doi.org/10.1016/j.energy.2014.02.069.

de Roo, G. and Parsons, J.E. (2011). A methodology for calculating the levelized cost of electricity in nuclear power systems with fuel recycling. *Energy Economics*, 33(5), 826–839.

Rothwell, G. (2015). *The Economics of Future Nuclear Power*. Routledge Press, Chicago, IL.

RTE (2021). Futurs énergétiques 2050 : les scénarios de mix de production à l'étude permettant d'atteindre la neutralité carbone à l'horizon 2050. Report, RTE, Paris.

Sainati, T., Locatelli, G., Smith, N.J. (2019). Project financing in nuclear new build, why not? The legal and regulatory barriers. *Energy Policy*, 129, 111–119.

Samadi, S. (2017). The Social costs of electricity generation – Categorising different types of costs and evaluating their respective relevance. *Energies*, 10, 356 [Online]. Available at: https://doi.org/10.3390/en10030356.

Scott, M. (2018). Nuclear power outlook. Report, Energy Information Administration, Washington, D.C.

SFEN (2017). Les coûts de production du parc nucléaire français. Report, Société Française d'Énergie Nucléaire, Paris.

SFEN (2018). Les coûts de production du nouveau nucléaire français. Report, Société Française d'Énergie Nucléaire, Paris.

Strbac, G. and Aunedi, M. (2016). Whole-system cost of variable renewables in future GB electricity system. Report, Imperial College, London.

Thomas, S. (2010). The economics of nuclear power: An update [Online]. Available at: http://www.psiru.org/sites/default/files/Thomas_economics.pdf.

Thurner, P., Mittermeier, L., Küchenhoff, H. (2014). How long does it take to build a nuclear power plant? A non-parametric event history approach with P-splines. *Energy Policy*, 70, 163–171 [Online]. Available at: doi:10.1016/j.enpol.2014.03.015.

Université de Chicago (2011). Analysis of GW-scale overnight capital costs. *Energy Policy Institute of Chicago Technical Paper*, 66, 450–461.

WNA (2020). Economics of nuclear power. Report, World Nuclear Association, London.

4

The Costs of Nuclear Fuel

Cécile EVANS[1], Sophie GABRIEL[2,3], Christian POLAK[4] and Philippe VALBUENA[1]

[1] *Orano, Châtillon, France*

[2] *I-tésé (CEA), Gif-sur-Yvette, France*

[3] *Université Paris-Saclay, Saint-Aubin, France*

[4] *Orano Mining, Paris, France*

The fuel cycle comprises activities not only relating to the supply of nuclear fuel to power plants, but also to the management of discharged fuel after reactor irradiation. It begins with the exploitation of uranium mines and ends with the storage of radioactive waste generated during reactor irradiation. It includes uranium production, conversion, enrichment, and fuel fabrication, all stages corresponding to the front end of the cycle. After reactor irradiation for a period of 4–5 years, the so-called spent fuel is cooled down in the reactor pool. Once cooled in the reactor pool, (depending on the spent fuel management options, detailed in Chapter 1, Volume 2), the fuel follows a certain number of stages until the radioactive waste is conditioned and stored in a deep geological layer. These stages constitute the back end of the cycle.

The following factors are important to consider:

– The content of spent fuel is radioactive and radioactive decay takes more or less time depending on the substances contained in the spent fuel, ranging from a

Nuclear Economy 1,
coordinated by Jacques PERCEBOIS and Nicolas THIOLLIÈRE.

few hundred years or less for short-lived products, to many millennia for long-lived products.

– Spent fuel contains fissile materials that can be reused in the reactor: the uranium contained in the irradiated fuel, known as reprocessed uranium, has a fissile content as high or even higher than natural uranium; plutonium is made up of isotopes, some of which are fissile.

During reactor irradiation, certain fertile materials can become fissile by neutron capture. In the water reactors used in our days, irradiation leads to a consumption of fissile materials. Different reactor designs – referred to as fast or Generation IV reactors – can produce more fissile material than they consume from fertile material.

The design and control of geological storage facilities for isolating radioactive waste strongly depends on the nature of the waste and its radioactive decay during stockpiling.

Two main basic options can be defined:

– Open cycle: the whole spent fuel is considered as waste and will be packaged for geological storage including transport operations, long-term storage, spent fuel conditioning and storage.

– Closed cycle: the spent fuel is processed, separating recoverable materials, which are recycled via the fabrication of new nuclear fuels. The metal structure waste and fission products are packaged in waste containers for geological storage, integrating various stages (transport, stocking, waste treatment and storage). Materials recycling enabling the supply of new MOX and URE fuels for reactor irradiation.

Chapter 1 in Volume 2 details the economics of the different cycle options.

The three phases of the fuel cycle are illustrated in Figure 4.1 with a focus on the temporal dimension for apprehending the particularity of fuel costs.

Even if the entire nuclear fuel cycle (in particular, a spent fuel management program) is implemented in stages, all the steps and their interfaces should be considered holistically when evaluating the overall technical and economic performance.

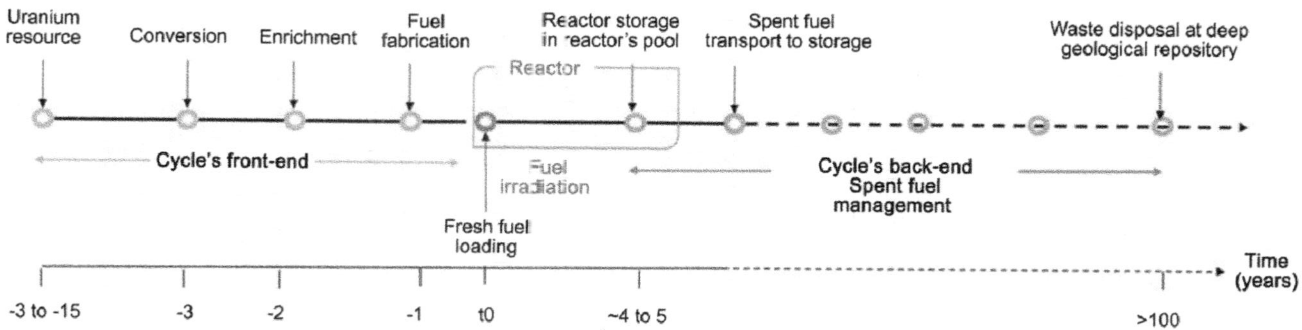

Figure 4.1. *Schematic view of the nuclear fuel cycle with a focus on three distinct phases. For a color version of this figure, see www.iste.co.uk/percebois/nuclear1.zip*

4.1. The cost of fuel expressed in the LCOE

The cost of fuel in the calculation of the levelized cost of electricity (LCOE) acknowledges the equilibrium valuation of various fuel costs, discounted on the date when the fuel is introduced in the reactor, and referenced to the discounted electricity production of this fuel (OECD/NEA 1993, 2013). The cost includes the purchase of materials and services, inventory financing – in particular natural uranium to ensure supply security – provisions for future charges associated with the management of spent fuel over a long period, beyond the reactors' operating period (see Figure 4.2).

Provisions for the future management of spent fuel are fully or partly covered by dedicated assets as in France, or by a contribution making it possible to constitute the dedicated fund managed by the electricity supplier or the entity in charge of managing radioactive waste (see Chapter 1, Volume 2).

The discount rate(s) used considering the time horizon differences of the various costs involved in the fuel cost have a significant impact on this theoretical valuation at equilibrium of the consolidated price of electricity. This approach with the rates often used divides the amounts of spent fuel management operations taking place several tens, or even hundreds, of years after nuclear production.

The breakdown of fuel costs according to this approach is illustrated in Figure 4.3, inspired in the representative analysis of the US market (World Nuclear Association 2017). The contribution for the management of spent fuel for existing US reactors was set at 0.1 cent/kWh by the "Nuclear Waste Policy Act of 1982".

According to this simplified approach, electricity suppliers consider that the fuel costs mainly depend on two factors: the price of components at the front end of the cycle (uranium as a resource, its conversion, enrichment, design and fuel fabrication) and in-core fuel management optimization.

As detailed below, the component costs at the front end of the cycle respond to markets regulated by the supply/demand dynamics. In-core fuel management optimization determines the quantity of nuclear material to be used to attain the nuclear production targets. It depends on two factors, the initial enrichment of the material and the fuel's maximum combustion rate, which are important fuel design parameters (World Nuclear Association 2017; Nuclear Energy Institute 2019).

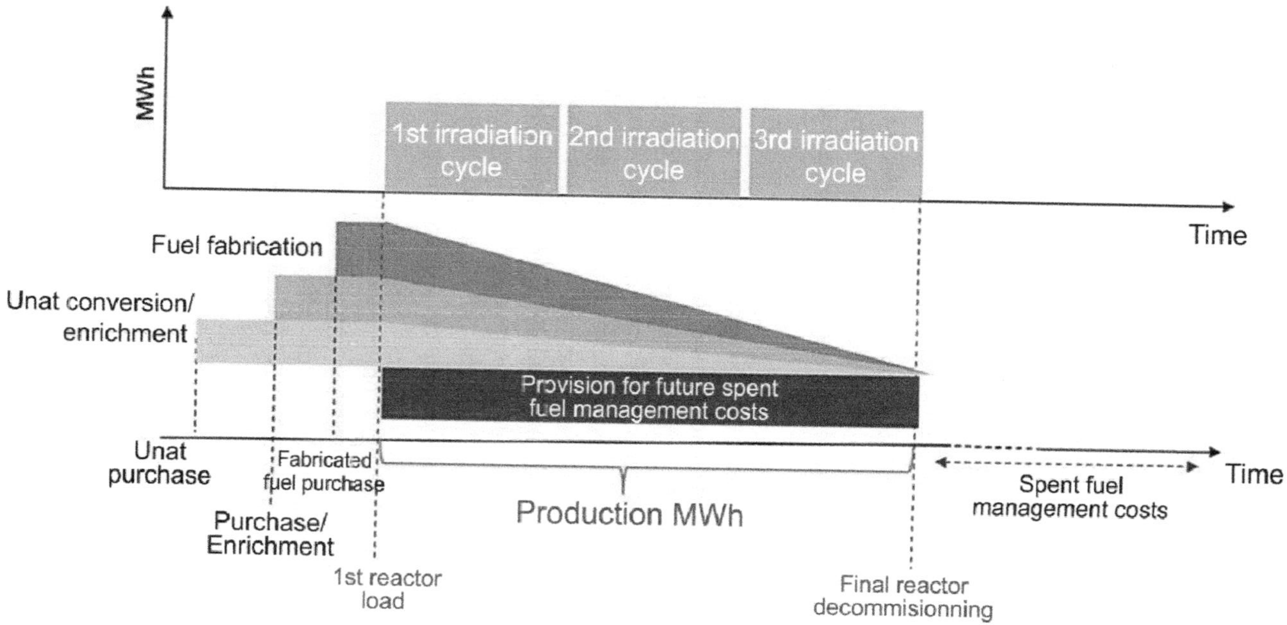

Figure 4.2. *Illustration of the fuel cost calculation methodology. For a color version of this figure, see www.iste.co.uk/percebois/nuclear1.zip*

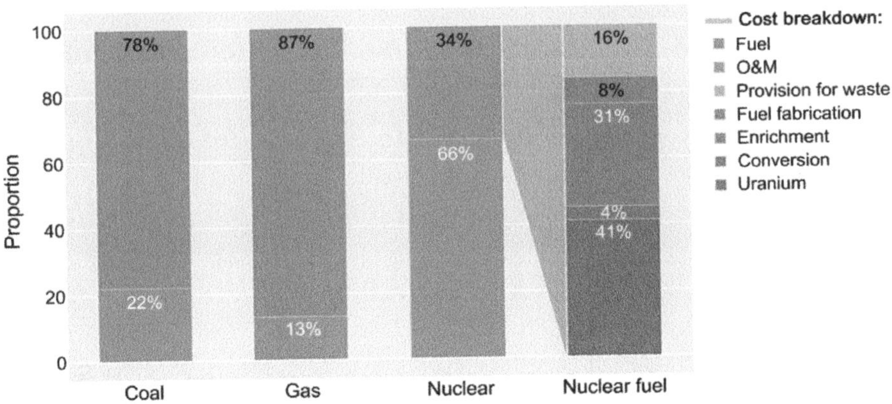

Figure 4.3. *Breakdown of operating costs for coal, gas and nuclear power (World Nuclear Association 2017) (source: Nuclear Energy Institute). For a color version of this figure, see www.iste.co.uk/percebois/nuclear1.zip*

The increase in combustion rates, possible thanks to the improvement in fuel design since the start of civil nuclear production, and the extension of the irradiation cycle duration have led to an optimum in fuel management costs, integrating the fuel cost and the maintenance shutdown costs during the intercycles. This optimization is limited by the current restrictions on the maximum tolerated ^{235}U enrichment level (set at 5%) and the maximum irradiation rate (62 GWd/t). These values vary from country to country, depending on the safety criteria imposed by the national safety authorities on the maximum fuel combustion rate for harnessing cycle times (variations between less than 12 months for Japan and 24 months for the United States) and the chosen cycle option (open cycle or recycling cycle).

The impact of these changes on the cycle's back-end operations and the corresponding costs must also be taken into account. The financial mechanisms for constituting back-end funds as well as the entity bearing the financial responsibility for such costs lead to different economic assessments, depending on the country and the stakeholders.

Take for example the contrasting dynamic in the United States, between existing reactors (for which the entire back-end financial burden is borne by the Department of Energy for a fixed contribution per kWh via the implementation of the "Standard Contract for Disposal of SNF and/or HLW") and the revision via the "New Standard contract" for new reactors, in which part of the financial responsibility is borne by the electricity supplier.

The recent OECD NEA report (IEA/NEA 2020) considers cycle costs over US$9/MWh for most countries.

One should note that the fuel cost structure varies over time, together with the progress of the spent fuel and radioactive waste management program and the residual operating time of power plants before shutdown, which can lead to significant variations in the contribution to cover the costs of the cycle's back end. Theoretical evaluations of the system idealized via the LCOE methodology do not make it possible to take such temporal evolutions into account. For this reason, more advanced valuation methodologies have been developed in order to capture all the stages of the fuel cycle, as well as the specific context of each country/each electricity supplier (see Chapter 1, Volume 2). Sweden, a country having opted for the open cycle and ahead in the implementation of its spent fuel management program and deep geological storage project, has revised this contribution upwards during the last 3-year assessments. In 2021, the back-end cost of the cycle corresponded to more than 25% of the production costs of nuclear electricity (SKB 2021).

Spent fuel management costs are the subject of significant attention in Sweden. The goal is to optimize the scheduled stages for conditioning spent fuel in waste packages and to place them in geological storage, so as to minimize costs via a reduction in the number of waste packages and the footprint of deep geological storage.

The following sections detail the front-end components of the cycle. The management of spent fuel and associated waste is described in Chapter 1 (Volume 2).

4.2. Uranium: availability and markets

Uranium, a metal with unique characteristics, is intended for only one use: the production of nuclear electricity fuel. This section will characterize its consumption, quantification (resources and reserves), exploitation, actors and production. The geological aspects of uranium are discussed in the Appendix.

4.2.1. *Uranium consumption*

Natural uranium contains 0.711% isotope 235 (^{235}U), the only fissile natural isotope in thermal neutron power reactors. A few power reactors can burn natural uranium (Candu reactor), but most of them require enrichment up to 5% in ^{235}U.

Small (research or naval propulsion) reactors may require a much higher enrichment, even over 90%. But those will not be discussed here because they do not influence the world uranium market.

During the isotopic enrichment process, uranium is separated into two quantities: highly enriched uranium (HEU) or low enriched uranium (LEU), intended for fuel fabrication, and on the other hand, depleted uranium (also known as uranium tails). The depleted residual material is not considered as waste because it can be used in a fast neutron reactor (FNR), a technology assumed to be available in the more or less long-term. FNRs will not be discussed further here because they do not consume natural uranium, but rather plutonium and depleted uranium. Below a 20% enrichment, the IAEA considers that the material is non-proliferating and enters the LEU category, whereas above 20% it is considered as HEU.

Consumption during the first year of operation is higher because a full reactor core must be loaded. The annual consumption is variable because it depends on the reactor's technology (Candu, BWR, PWR, etc.), fuel and management (^{235}U enrichment of new assemblies, unloading combustion rate, possible recycling of fissile materials: uranium oxide [UOX], mixed uranium–plutonium oxide [MOX], or enriched reprocessed uranium [URE]) and the economic conditions of the world markets for uranium and enrichment services.

The average annual consumption of a reactor is between 150 and 200 tUnat/GWe/year.

The amount of natural uranium required to produce fuel is determined by the ^{235}U enrichment level required and the uranium tails assays. To some extent, it is possible to replace natural uranium with enrichment services and vice versa. If the price of uranium increases, it may be interesting to use less uranium and to enrich more of it (provided that the enrichment cost does not also increase in the same proportions. There is an economic optimum between the amount of natural uranium consumed and the amount of enrichment services required).

Global uranium consumption is related to installed nuclear capacities; in 2018, it was 59,200 tU for a world fleet of 396 GWe. The breakdown by geographical area is shown in Figure 4.4 (NEA-IAEA 2020).

Based on the declarations of the States having taken part in the NEA-IAEA publication, the forecasts for the fleet's evolution until 2040, and the associated consumption, are given for two scenarios, high and low (Figures 4.5 and 4.6).

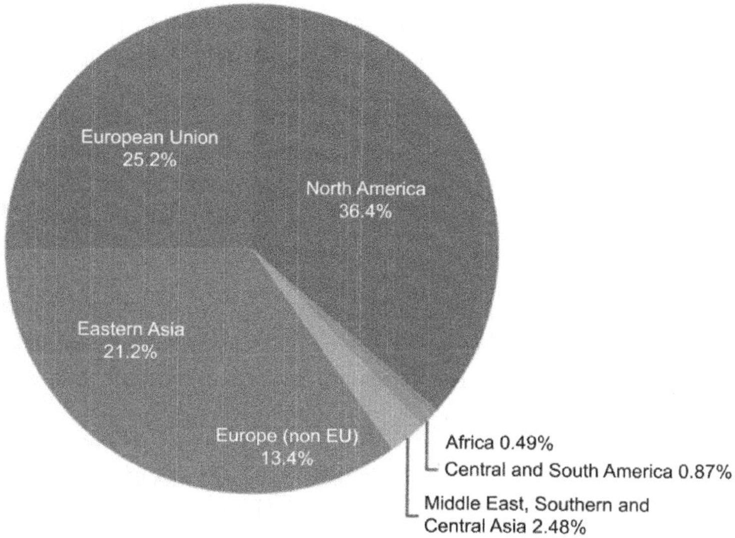

Figure 4.4. *World demand for uranium in 2018 (59,200 tU) (source: NEA-IAEA 2020; Red Book). For a color version of this figure, see www.iste.co.uk/percebois/nuclear1.zip*

The global or national evolution of uranium consumption over the longer-term is not detailed here because it is more related to the planned evolution of the nuclear fleet, as described in the chapter dedicated to the prospective scenarios by 2050 (section 5.1, Volume 2). Let us specify that to estimate this world consumption, it is necessary to take into account not only the need for the reactors themselves, but also the potential of secondary resources. Information should include a number of the reactors in place and those put into service, their technology (in particular thermal or fast neutron), their load factor, the possibility of recycling their fuel and the conditions of use of the fuel cycle (the tails assays of enrichment facilities, for example).

4.2.2. *Terminology, resources and reserves*

4.2.2.1. *System history and comparison*

The ultimate interest of uranium ore deposits is to be transformed by mining, with resources defined on a geological basis and reserves defined on an economic basis. Resources and reserves are critical elements for obtaining confidence for the miner, the financier and the public authorities in charge of issuing the exploitation permit.

The mining industry has made great progress in calculating resources with better quality data and increasingly sophisticated models, such as 3D imaging.

The mining industry is international in nature, often a mining company works on several continents. At the beginning of the 20th century, national standards emerged, which later became widespread in view of obtaining a common terminology. Miners were thus obliged to leave their national backyard for the international market and several sites were developed.

The terminology as we know it was constructed in the 1950s–1970s. The Australian Joint Ore Reserves Committee Code (JORC) was established in 1971. Its members were advised to standardize reports, engage competent and committed staff and fix a common terminology.

The changes in the mining industry implied that large companies gave way to small exploration companies, called Junior Mining. These were granted access to the market, but the Stock Exchange required certain discipline. The JORC was thus integrated within the Australian Stock Exchange (ASX) in 1990.

The mining industry dominated by Canada and Australia has imposed codes applicable to mining (the National Instrument NI 43-101 for Canada, and the JORC), which are also shared by investment funds.

The CRIRSCO (Committee for Mineral Reserves International Reporting Standards), a supranational committee created in 1994, has also drawn up a code that aims to be more universal, but which is nonetheless heavily inspired in the JORC's know-how. In 1997, the committees from the United States, the United Kingdom, Canada, Australia and South Africa agreed to work within the CRIRSCO framework and a model was published in 2006. The Russians joined in 2010 and Kazakhstan in 2014. The Chinese adopted the JORC for their uranium mines outside China and translated CRIRSCO into Chinese in 2014. Equivalences were formalized between the main countries. But the CRIRSCO appeared to be more supranational than practical: the major world mining exchanges all adopted the JORC and the NI43-101 in the early 2010s.

The JORC and the NI43-101 are essential tools for the mining industry with a three-pillar base: data transparency, competence and materiality, all sealed by a competent person with at least 5 years of experience. Not only do these codes converge on definitions, but also on a shared methodology. Only competent persons, who engage their criminal liability, are authorized to sign these resources and reserves.

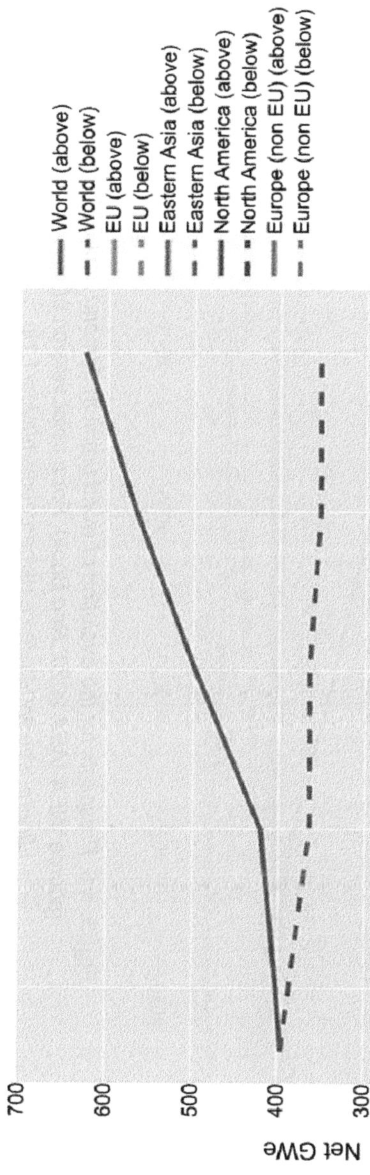

Figure 4.5. *Prospects for the development of the world's nuclear power plants until 2040 (source: NEA-IAEA 2020; Red Book). For a color version of this figure, see www.iste.co.uk/percebois/nuclear1.zip*

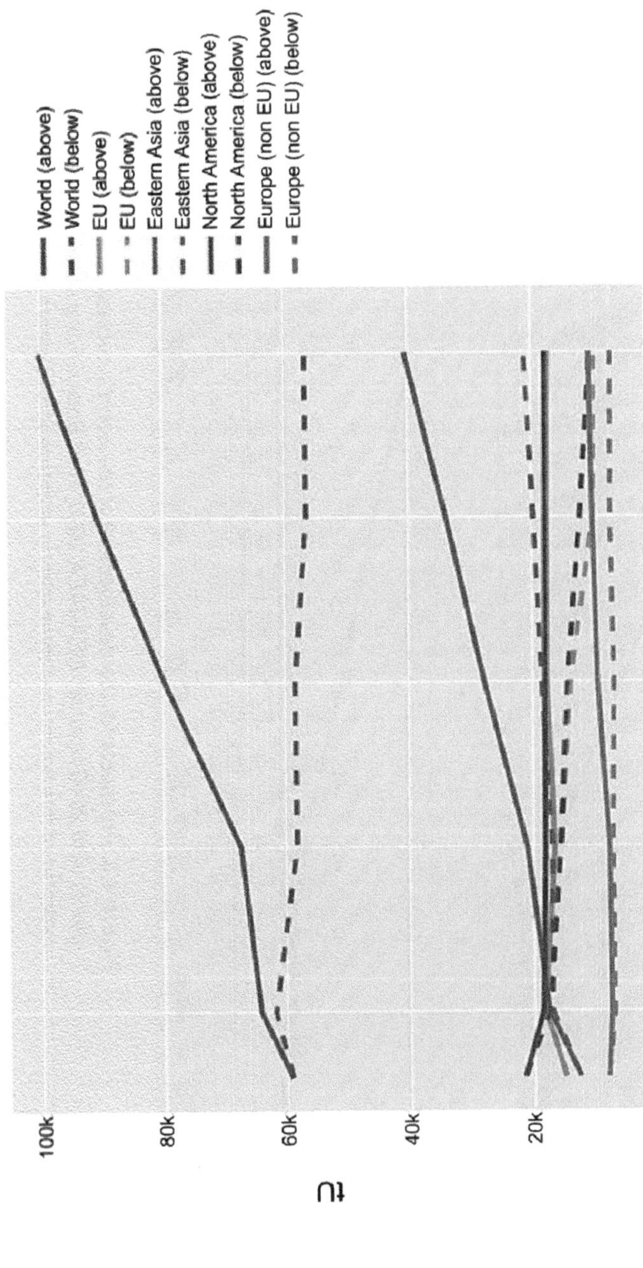

Figure 4.6. *Prospects for the evolution of world uranium demand until 2040 (source: NEA-IAEA 2020; Red Book). For a color version of this figure, see www.iste.co.uk/percebois/nuclear1.zip*

4.2.2.2. *Terminology*

4.2.2.2.1. Resources and reserves

Exploration provides convincing elements to define a target, a prospect for a mineral deposit in a geological context. While a blank target does not have any resources, a program is needed to set goals.

A resource is a mineral concentration characterized by a certain quantity (tonnage) and quality (grade). The elements are derived from exploration work, sampling, surveying and trenching (or box-cutting) so as to define the resource's quality over time and improve its status: inferred, indicated and measured. Resources can be classified as follows:

– Inferred resources: there is enough geological information for a first tonnage and grade estimate, but without guaranteeing grade continuity and geological quality. There is an implicit continuity.

– Indicated resources: this category provides enough information on the basis of the work deemed necessary to envision a model with continuity and geological understanding, as well as to establish a mining plan and economic outlook. There is a supposed continuity.

– Measured resources: the level of confidence of all the data generated by the density and quality of the works, reinforcing the zone's continuity, makes it possible to initiate a detailed mining plan for economic assessment and viability. There is a confirmed continuity.

The notion of continuity is understood in terms of size, mineral and metallurgic appraisal of the entire ore deposit. The geologist's solid knowledge as to the interpretation of the geological model and coherent statistical data make it possible to validate these resources. The big fault in resource estimate is an erroneous geological model with insufficient data.

A mineral reserve is defined by the transformation of indicated and measured resources into a technically and economically exploitable entity on the basis of pre-feasibility or feasibility studies. Such studies demonstrate that, at the time of reporting, extraction could reasonably be justified. This conversion results in factors which reduce tonnage based on mining, metallurgical, environmental, legal, etc., considerations.

Probable reserves are derived from the most reliable indicated and proven resources and from the conversion of measured resources.

4.2.2.2.2. Framework studies

A scoping study (or preliminary economic assessment) can be based on inferred resources or a proven prospect, but in no way can it lead to the definition of reserves. It provides a reasonable technical and economic order of magnitude for a potential, with a ± 30–40% accuracy.

A pre-feasibility study (or preliminary study) provides possible options on the mining method and sequence, the metallurgy, profitability and makes it possible to turn resources into reserves for assumed financial conditions. A qualified person will be involved and the accuracy regarding reserves will be ± 20–25%.

Thanks to a feasibility study, one can choose the most favorable technical and economic option for the exploitation of the mine. It must provide information for supporting a financial investment decision according to a detailed production schedule. Accuracy on reserves varies between 10% and 15%.

4.2.2.3. *Resource classification according to the IAEA and the NEA: the Red Book*

For uranium, a group of Member States within the IAEA (International Atomic Energy Agency) and the NEA (Nuclear Energy Agency) has been using a specific and original classification based on production cost tranches since 1965.

The results are published every 2 years, in the Red Book, by the OECD General Secretariat. The 28th edition (2020) collected information from 45 countries[1]. An entire chapter is devoted to uranium supply, cross-referencing information by country, depending on resource quality, processes and industrial locations. Domestic and foreign exploration budgets are also reported. Section 4.2.1 presents consumption by country and a scenario until 2040. Section 4.2.5.2 covers uranium and nuclear activity by country.

The Red Book presents identified resources. They include reasonably assured resources (RAR) and inferred resources. RAR cover indicated and measured reserves and resources according to the definition used by the mining industry. Inferred resources also correspond to those from the mining industry. However, these identified resources are classified into four production cost tranches, estimated in US$/kgU. Table 4.1 presents the cost tranches considered by the IAEA-NEA. Commercial prices being expressed in US$/lb U_3O_8, the second row of the table presents the equivalence, bearing in mind that 1 kgU = 2.6 lbU_3O_8.

1 Available at: https://www.oecd-nea.org/upload/docs/application/pdf/2020-12/7555_uranium_resources_production_and_demand_2020__web.pdf.

	Tranches			
US$/kgU	<40	<80	<130	<260
US$/lbU$_3O_8$	<15	<31	<50	<100

Table 4.1. *Cost tranches considered by the IAEA-NEA*

This classification by cost is unique and identifies major categories – in particular ISR (in situ recovery) deposits – as well as inconsistencies in the first two tranches (Figure 4.7).

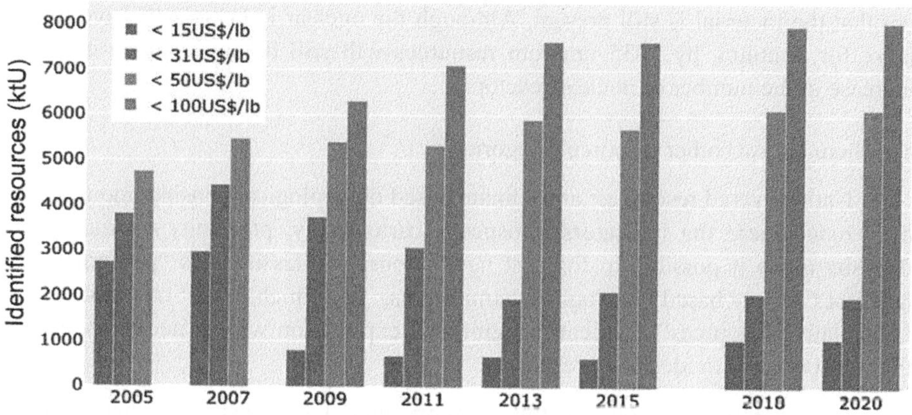

Figure 4.7. *Evolution of resources identified by cost tranche (source: Orano according to the Red Book). For a color version of this figure, see www.iste.co.uk/percebois/nuclear1.zip*

However, it should be emphasized that the classification of deposits in this system amounts to the responsibility of States, and it is rarely supported by consultable studies. In many cases, these resources lack the computational rigor found in the industry. Some countries transpose the declarations of their miners, whereas other countries make their own estimates, though without ensuring the three pillars of the JORC or NI43-101 classification (transparency, competence and materiality).

The advantage of this classification is that it offers an overview and identifies the major trends in cost evolution, something which the Red Book group takes into account. The US$26/kgU category disappeared in the early 2000s and the

US$260/kgU category was introduced in 2009, confirming an increase in costs in the mining industry in general.

We obtain a total (<260 USD/kg) of 8.078 MtU of resources identified in 2019, that is, 1.3 MtU more over 10 years, which represents a considerable effort.

A new category, in situ resources was introduced in 2016, assuming fully recovered underground uranium, without taking mining and metallurgical yields into account.

The average cost practically doubled between 2005 and 2008, today being close to 70 US$/lb U_3O_8. We can clearly see that low-cost deposits are becoming scarce, but that the material is still present. Although the nuclear industry will have to pay more for uranium, by 2035 uranium resources will still be sufficient to meet the increase in the number of nuclear reactors.

There are two other resource categories:

– Undiscovered resources: an estimate based on geological considerations, either in a zone where the indicators, prospects, cartography, proximity of known ore deposits make it possible to forecast new resources (classified as “prognosticated resources”), or based geological similarities, in which case one speaks of “speculative resources”. Evidently, significant exploration will be necessary to turn these resources into identified resources.

– Unconventional resources: this category includes deposits of other minerals (phosphates, copper, etc.) where uranium is present in small, minor or even incidental quantities, but which can be specifically recovered during the main treatment and lead to notable productions. In that case, one speaks of a co-product when its value is decisive for the general cost or a by-product when it is a complement.

4.2.3. *Uranium resources*

4.2.3.1. *Conventional or classic resources*

The tools developed to calculate resources have considerably improved and homogenized over the past 50 years. With repeated exploration efforts, but closely related to the price of uranium, the progression of available resources, clearly characterized by the mining industry, has undergone significant upheavals (see Figure 4.8).

There was a clear growth in resources between 1965 and 1985 correlated with the massive arrival of nuclear energy. These resources became stable over the 1985–1995 period, and decreased in 2005 as a result of the destocking of weapon-related matters between 1993 and 2013, with a subsequent fall in prices (Russian-American disarmament agreements). The strong recovery observed since 2005 can be explained by the nuclear power revival, until the advent of the Fukushima accident. Today, there are 4.7 MtU of RAR.

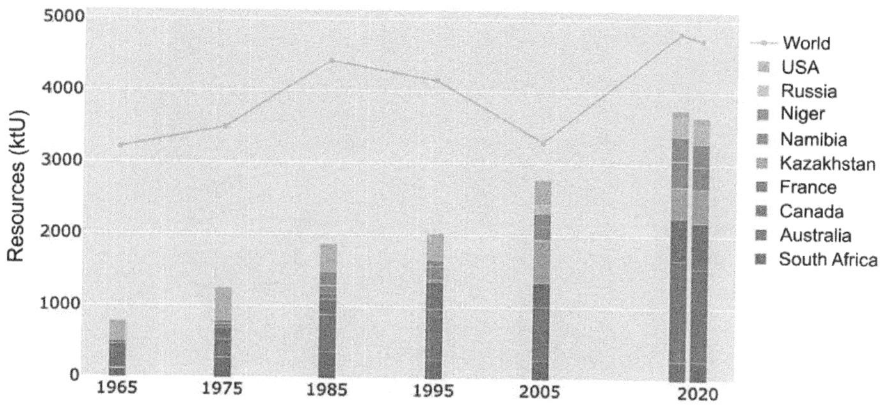

Figure 4.8. *Evolution of reasonably assured resources by country (source: Orano, compiled from several Red Book editions). For a color version of this figure, see www.iste.co.uk/percebois/nuclear1.zip*

For the year 2018, we can observe the importance of Australian resources with Olympic Dam (75% of the country's RAR), and the increase of those in Canada (discoveries of Arrow, Triple R, Wheeler River, Shea Creek), Niger with the Imouraren deposit and Namibia with the Husab. Although the resources in the United States have fallen sharply, they are better qualified.

In South Africa, uranium is a by-product of gold deposits. As the costs of gold mining are the highest in the world, this country's uranium resource becomes unlikely.

Considering all the identified resources (RAR + inferred), the 2019 total was 8.01 MtU; please refer to Figure 4.9 for further details.

Countries with over 300 ktU resources will continue to act as major stakeholders for a very long time. It should be noted that the two leading world powers, China

(290 ktU) and the United States (100 ktU), have limited resources and their nuclear fleet is dependent on imports.

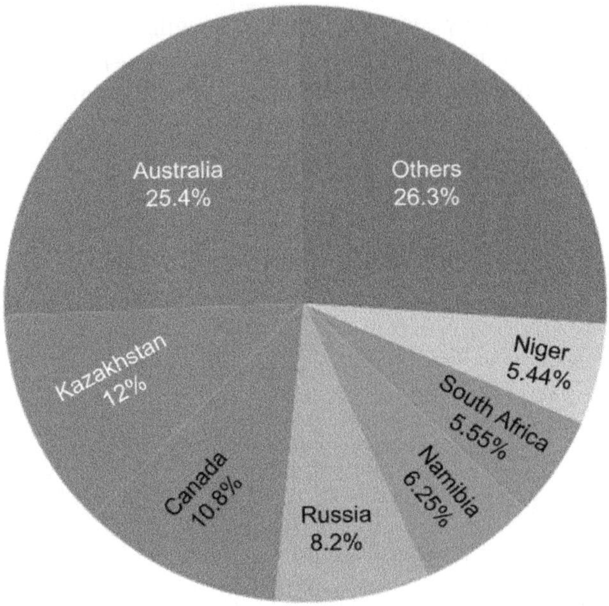

Figure 4.9. *Resources identified by country in 2020 (source: Orano according to the Red Book). For a color version of this figure, see www.iste.co.uk/percebois/nuclear1.zip*

A more mining-oriented interpretation of these resources can be achieved by working with the compliant data from the JORC and NI43-101 codes, reported by the mining companies. Orano has been compiling data from these companies since 2004, with the majority based on JORC or NI43-101 codes, for deposits over 100 Mlbs U_3O_8 (that is, 38 ktU) and between 50 and 100 Mlbs. Data compilation amounts to a total of 6.8 MtU resources (among measured, indicated and inferred resources). The 1 MtU difference with IAEA corresponds to the deposits from countries that have not yet brought their resources into conformity (notably, India, Brazil and China).

We can therefore conclude that these resources published by the IAEA-NEA and those from mining companies provide reliable resources, which represent 100 years of current consumption. However, the extraction cost will be much higher than the current price of uranium.

4.2.3.2. *Non-classical resources*

4.2.3.2.1. Uranium from phosphates

The IAEA-NEA classification has created a specific category for unconventional resources, specifically addressing uranium recovery during the treatment of phosphate ore. This extraction method is well known, and between 1978 and 1998 the United States produced about 18 ktU with a process developed by the Oak Ridge National Laboratory (ORNL). All their units located in Florida and Louisiana closed when the price of uranium sharply fell after the Chernobyl accident, but especially with the marketing of Russian military stocks (Megatons to Megawatts agreement 1993–2013), which kept prices at surprisingly low levels.

The ORNL process is a liquid–liquid extraction of uranium from phosphoric acid produced by the sulfuric attack of the phosphate ore.

Research on extractants continues to reduce costs and extraction loops. But fundamentally the economy is dependent on the uranium configuration of the phosphates mined, which varies from country to country – and even within the same deposit – in proportions of 50–250 ppm.

Some phosphate ore deposits exceed 0.1% U_3O_8 (Brazil, Central African Republic, Congo, Colombia), in which case uranium becomes a product which has to be extracted so as not to contaminate the phosphate.

What can be expected from this uranium production? By definition, it is inelastic because it depends on phosphate production (and we will not produce more phosphates to produce more uranium). Relying on these unconventional resources following the IAEA definition makes really no sense. In addition, they have only rarely been calculated according to mining rules, but most often estimated on the basis of scattered analyses.

Another approach would be to consider existing phosphate production flows. The sedimentary deposits of the Middle East, the Mediterranean rim and the Atlantic coasts are the most likely to be treated, with approximately 85 Mt/year of phosphates having an average U_30_8 grade of 100 ppm. Assuming an exhaustive extraction of uranium, approximately 8,000 tU/year would be produced, that is, less than 15% of the current world production. This is not negligible, but assumes full recovery, in other words, a lot of facilities.

4.2.3.2.2. Uranium from black shales

Black shales are very abundant on the Earth's surface and extend over large areas. They correspond to sedimentary formations developed in anoxic conditions, and sometimes having metamorphosed. They contain many interesting metals (Cu, Co, Ni, V, Mo, Re and U) but rarely in economic amounts, except for copper in Poland (Kupferschieffer), in the Copper Belt of Katanga, in Zambia (associated with cobalt) and in Finland (associated with nickel and cobalt). A few attempts at the specific extraction of uranium (Sweden, Estonia) have failed to show an economic interest. The processes are challenging and costly, with insufficient uranium concentrations – between 25 and 250 ppm – to justify an extraction. High uranium prices and co-extraction with innovative processes (bioleaching) could justify the recovery of minor metals such as uranium.

The resources are potentially very important. In central Sweden, after specific exploration work, a real resource has been estimated and this only on a small area of the Storsjön black shale formation. Two deposits, Haggan and Viken, yielded 800 ktU of inferred resources with U_3O_8 grades of 150–170 ppm, according to the JORC code. At present, the project is oriented toward the recovery of contained vanadium and molybdenum.

This means that there is an enormous European resource, but at a high cost. The extraction of 200–250 tU/year associated with Cu-Co-Ni from a metamorphosed black shale in Finland (TerraFame company) could encourage Europe to review its supply strategy by securing low-carbon energy.

4.2.3.2.3. Uranium extraction from seawater

The concentration of uranium in seawater is very low, around 3 µg/L, but the volume of the oceans is such that this still represents a quantity of dissolved uranium of 4.5 billion tons, compared to the 17 MtU of the current conventional terrestrial resource (IAEA-NEA). For a number of reasons, the extraction of significant amounts of uranium from seawater today remains more of a dream than a reality. To begin with, pumping seawater to extract this uranium would require more energy than the one produced with recovered uranium. If one intends to use the existing industrial throughputs, for example, on a nuclear power plant, it appears that the possible annual quantity is extremely low. In fact, huge amounts of water need to be processed. To produce the world's annual consumption of uranium (about 65 ktU), it would be necessary to extract at least all the uranium from 2×10^{13} tons of seawater, the equivalent volume to the entire North Sea. In fact, it is only the great ocean currents that provide these enormous quantities without pumping, and the goal is to try to extract this uranium even partially.

Numerous research works have been published on the study of adsorbents immersed in these major currents. After immersion, the adsorbents are chemically treated in order to recover the uranium. The quantities extracted are still low in view of the complex and costly operations to be carried out at sea. One kilogram of adsorbent, after 1 month of immersion, yields a few grams of uranium and the adsorbent can only be recycled a few times due to the loss of efficiency. An industrial extrapolation exercise carried out for the extraction of 1,200 tU/year – the order of magnitude of a mine's capacity – results in a very expensive facility, requiring over 1,000 km^2 at sea, an armada of boats for transport and whose operation poses significant ecology-related issues.

Current research aims to increase the efficiency of adsorbents, but it is clear that nowadays it would be very risky to develop a long-term industrial strategy based on a large production of uranium from seawater at an affordable cost.

4.2.3.2.4. Coal ash uranium

Coal reserves contain uranium at highly variable concentrations. In the past, several facilities have been able to extract uranium from coal ash. For example, during the 1960s-1970s in the United States, 1,100 tU were extracted by this method. This figure can be compared to the amount of uranium produced from phosphate rock over the same period: 17 ktU in the United States.

The reason coal ore is not processed directly to recover uranium is due to uranium reconcentration during coal combustion. Actually, almost all of the uranium is found in the solid residues from combustion (the ashes).

The recovery of uranium from coal ash can be considered as the extraction of a by-product or a co-product. The distinction can be made on the uranium content of the ore. Uranium-rich ore deposits (grade of the order of magnitude of other so-called unconventional deposits: 200–300 ppm and over) could lead to the production of a co-product, or even a primary product from uranium. Low-grade deposits eventually give rise to the production of by-products, coal production being profitable on its own.

Coal resources and reserves are carefully inventoried at a global scale. However, this inventory does not provide information about the coal's composition. The identified or inferred resource can be estimated at technically extractable 1.1 and 4.5 MtU. However, only a maximum of 200 ktU could be turned into reserves in the 21st century.

Despite this, the production capacities of uranium from coal ash on a large scale are constrained by the consumption of coal and lignite in the energy sector. For a very optimistic scenario regarding the available fraction of the ash and its content (all the available ash should be processed, without any limit grade), uranium production could reach 9 ktU/year. By imposing a cut-off grade constraint of 200 ppm to the ashes, the coal ash treatment does not produce over 700 tU/year, that is to say, around only 1% of current annual needs.

The potential reserves of uranium from coal ashes thus appear negligible compared to other sources of uranium and to the natural uranium needs in the long-run. Nevertheless, the extraction of uranium from coal ash is technically and economically feasible in situations such as a particularly uranium-rich coal ore deposit, and can therefore act as a supply source for a local uranium need.

4.2.4. *Secondary resources*

Secondary uranium resources are fissile material resources that have already been mined; not only do they include the various commercial or governmental stocks, in different forms (natural, depleted or enriched uranium), but also the materials resulting from the recycling of spent fuel (reprocessed uranium and plutonium).

4.2.4.1. *Commercial and government stocks*

Mining production has been far greater than the electricity demand. According to the information gathered by the NEA, the cumulative production until 2012 was estimated at more than 2.7 MtU, whereas the cumulative reactor demand was estimated at 2.15 MtU. These figures give an approximation of the upper limit of what could eventually be available for the commercial sector: 550 ktU.

The surplus from mining production made it possible to stock fissile material primarily intended for military uses. While part of the stock has been used as fuel in certain reactors and for nuclear weapons tests, the major portion has not yet been consumed and is found in the form of stocks of various kinds, essentially in the hands of the United States and Russia. These stocks are not necessarily "secondary resources"; for the military stocks of fissile materials to be considered as secondary resources, recycling has to be contemplated for power reactors.

Civil or commercial stocks include strategic stocks, pipeline stocks and excess stocks available on the market. They are owned by producers, consumers or governments. For confidentiality reasons, the information on the size of these stocks is strictly limited. Only a few countries have declared Unit (natural uranium) stocks to the NEA, for example, Argentina (52 tUnat), South Korea (2,000 tUnat), Portugal (168 tUnat), Switzerland (1,543 tUnat) and the United States (34,375 tUnat)[2].

4.2.4.2. *Materials drawn from recycling processes*

The recycling of spent fuel from reactors depends on the treatment, enrichment and fuel fabrication capacities for MOX (mixed uranium and plutonium oxide) or URE (enriched reprocessed uranium), and from the reactors authorized to burn this type of fuel.

In the short and medium-term, there are no possible surprises, the flows being limited by the capacities of existing facilities and representing a saving of 2,500 t Unit/year, at best.

In the longer term, recycling depends on the countries' nuclear policies and the possible construction of new facilities. It should be noted that the new reactors are designed to be able to consume MOX, which is not the case for all past reactors. However, it is doubtful that treatment capacities will be developed for recycling in PWRs, but will rather be intended for FNR recycling, which changes the whole issue of secondary resources.

4.2.4.3. *Depleted uranium and underfeeding*

The amount of natural uranium required to produce an amount of enriched uranium is determined by the level of enrichment required and the tails assays. The ^{235}U grade of depleted uranium is variable and is determined by the enricher, depending on the uranium market conditions and the enrichment services. Uapp stocks related to past enrichment activities contain more or less ^{235}U according to the economic conditions at the time of their creation and may possibly be subject to further enrichment.

Current stocks of Uapp, which have real potential as secondary resources, result from the period when military arsenals were built and belong to the US and Russian governments. At present, Uapp re-enrichment is relatively easy in Russia due to their large enrichment capacities. On the contrary, such capacities are not available in the United States, where the uranium and enrichment markets are not to be disrupted.

An enricher has some flexibility in relation to the plant's operational tails assays. If the operational tails assays are lower than the contracted value, excess natural

2 US figures are published annually by the EIA (USA Energy Information Administration).

uranium can be set aside and sold on its own account (as natural or enriched uranium). This is known as underfeeding. The opposite situation, in which the tails assays are higher, requires the enricher to supplement the quantity of natural uranium supplied by the backer of the project. In both cases, the enricher will base its decision on the economics of the plant, as well as uranium and energy prices.

The secondary resource related to the current enrichment excess capacities (namely Uapp re-enrichment and underfeeding) can be estimated at about 5,000 tUnat/year.

In recent years, enrichment excess capacities have been strongly associated with the demand decrease following the Fukushima accident. However, the development of new capacities has been suspended while an increase in demand is expected. If, as predicted by many scenarios, there is an increase in the nuclear demand, there will be a match between the supply and demand of SWUs (separative work units).

4.2.5. *The uranium market*

4.2.5.1. *Uranium production*

The discovery of the mineral autunite (uranium phosphate) by M. Champeaux at Saint-Symphorien in Burgundy, reported in the *Annales des Mines* from 1801, was fortuitous!

Uranium has been produced since the middle of the 19th century. At first, it was done during the second half of the 19th century, for obtaining pigments in the form of uranates. Added in small quantities (<2%), these were used for manufacturing "ouraline" glass, with strange colors and fluorescence. The Austro-Hungarian Empire monopolized this production. The pitchblende (uraninite) from the Jachimov mine tailings, in Bohemia, was sufficient to supply a uranate production plant, providing around 2 tU/year.

The discovery of radioactivity changed things, enticing a few mining companies to embark on the adventure. In Cornwall, it was a collaborator of Marie Curie who had a meeting with an English Lord! In Colorado, where carnotite was reportedly known by the Ute natives to color their faces, production started in 1913 and threatened to break the European monopoly.

Then, with the discovery of a uranium deposit at Shinkolobwe in the Congo in the 1920s, and in Saskatchewan in the 1930s, these two countries shared the radium market, but not enriched uranium!

The production of uranium and its pigments accumulated before 1945 has been estimated at 7,500 tU (Figure 4.10).

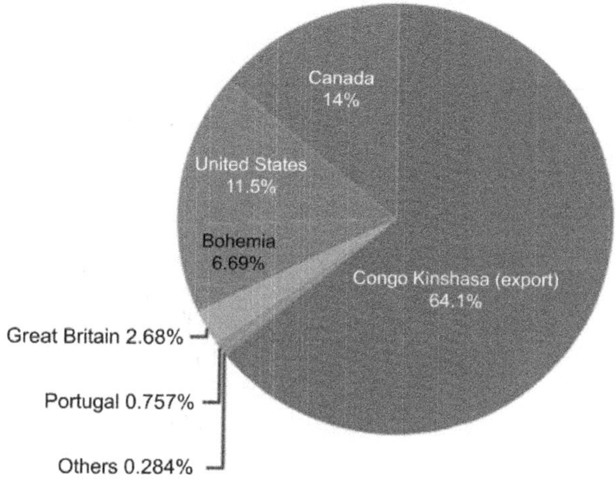

Figure 4.10. *World uranium production 1853–1944 (Orano estimate). For a color version of this figure, see www.iste.co.uk/percebois/nuclear1.zip*

After 1945, production underwent several periods, as shown in the production graph per country (Figure 4.11) in thousands of tU:

– 1945–1965: first wave, excessive production for military use. United States is the lead producer worldwide, followed by Canada. The USSR (mainly Central Asia) and its European satellites come in the third place.

– 1965–1990: second wave, the use of nuclear energy develops strongly with major programs (Europe, USA). Emergence of new producers: South Africa, Australia, Niger, Namibia.

– 1990–2005: the recess, since the end of the Cold War with the use of military stocks (from the first wave) and the effects of Chernobyl. While European and American producers tend to disappear, the dismemberment of the USSR brings about new stakeholders, Kazakhstan, Uzbekistan and Ukraine.

– 2005–2011: third wave, the nuclear revival until the Fukushima accident. Kazakhstan takes 40% of the market, followed by Canada and Australia.

– Since 2011, falling prices and the slowdown in nuclear programs provoke an increase in stocks, companies reducing their costs and production. Few countries continue to operate, where ISR technologies predominate (Figure 4.12).

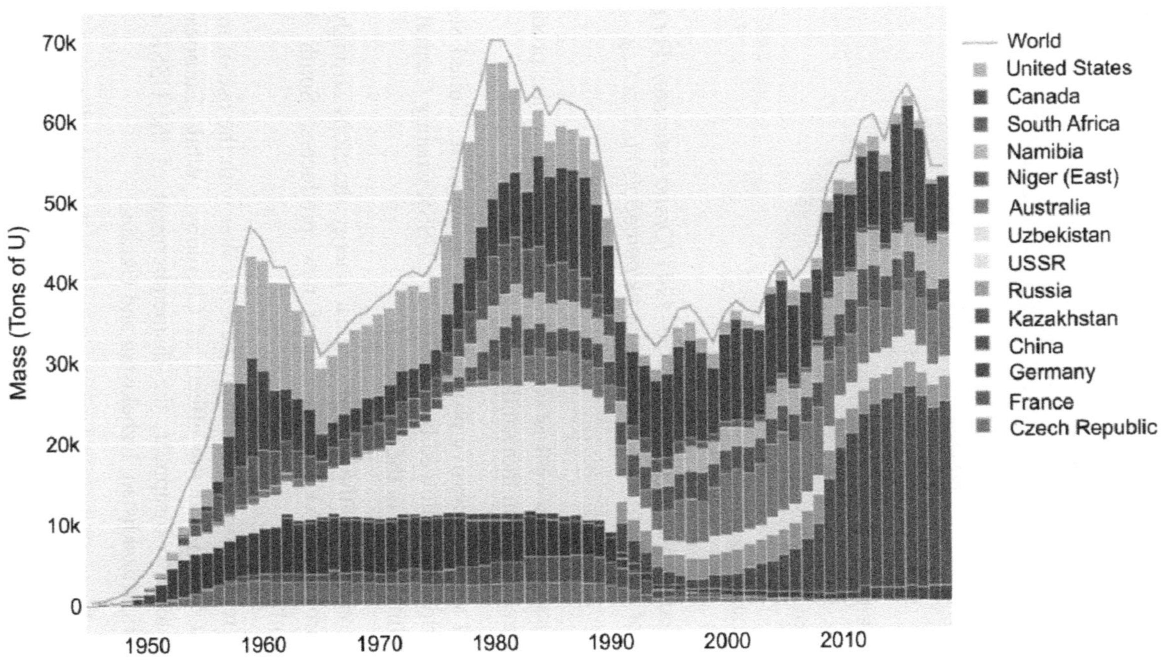

Figure 4.11. *Production by country. Only countries with a cumulative production greater than 1% of the total have been retained (source: IAEA-WNA). For a color version of this figure, see www.iste.co.uk/percebois/nuclear1.zip*

In 2019, six countries, Kazakhstan, Uzbekistan, Niger, Namibia, Canada and Australia, concentrated 87% of production for commercial purposes, taking part in international trade.

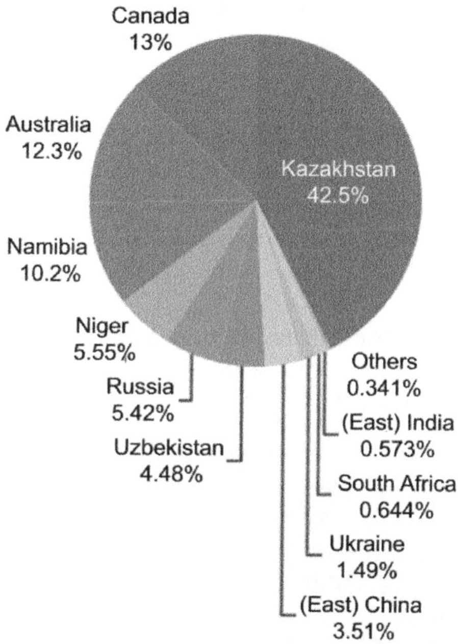

Figure 4.12. *Breakdown of world uranium production 2019: 53.718 tU (source: WNA). For a color version of this figure, see www.iste.co.uk/percebois/nuclear1.zip*

Since the end of the 20th century, the increase in the demand for certain minerals has led to the proliferation of major mining projects. Firms wishing to set up new mining operations are increasingly coming up against opposition from local populations supported by environmental NGOs. Conscious about sustainable development requirements and the stakes represented by their environmental image, some firms have succeeded in developing a strategy, a communication and the actions needed to make their mining projects more acceptable, even when these have a considerable impact on the environment.

Uranium mines have followed the same developments as in the mining industry. Nowadays, uranium extraction is performed using modern mining techniques, under regulatory control, with public participation in decision-making and in compliance with clear site restoration obligations, which provide for the rehabilitation of mines at the end of their operation.

4.2.5.2. *Market stakeholders*

The uranium market is characteristic for having national actors with an international vocation, as well as some mining actors on the stock exchange (Major or Junior Mining), dedicated to an ore deposit. This is the consequence of policies implemented in many countries due to strategic reasons and national independence concerns.

The dismemberment of the USSR generated three State actors actively involved in the market: Kazakhstan (1st world producer), Uzbekistan and Russia. Since then, the production in Eastern countries, which had been high during the Soviet era, particularly in East Germany (218 ktU) and Czechoslovakia (112 ktU), has completely ceased.

As it happens with countries, there are only a few major mining stakeholders (eight producers who concentrate 75% of the world's production), whose influence has varied in recent years (Table 4.2).

	2013	2019
Kazatomprom	25%	22%
ARMZ-U1	14%	13%
Orano	11%	10%
Cameco	15%	8%
CGN	1%	7%
BHP	6%	6%
CNNC	3%	4%
Rio Tinto	5%	3%
Paladin	5%	–
Others	16%	27%
World production in tU	60,065	53,718

Table 4.2. *Proportional breakdown of the world's production by mining company between 2013 and 2019*

Kazatomprom reduced its production to meet the market's needs. Chinese companies (CGN and CNNC) have increased their participation in foreign investment from 4% to 11%. The decline of Cameco and Orano is explained by the shutdown of McArthur mine due to market reasons in 2018. Paladin (Langer Heinrich ore deposit) ceased its production in 2018 due to economic reasons. Rio Tinto sold its stake in its Rössing mine to CNNC, thus preventing its disappearance by the end of 2020 with the shutdown of the Australian Ranger mine.

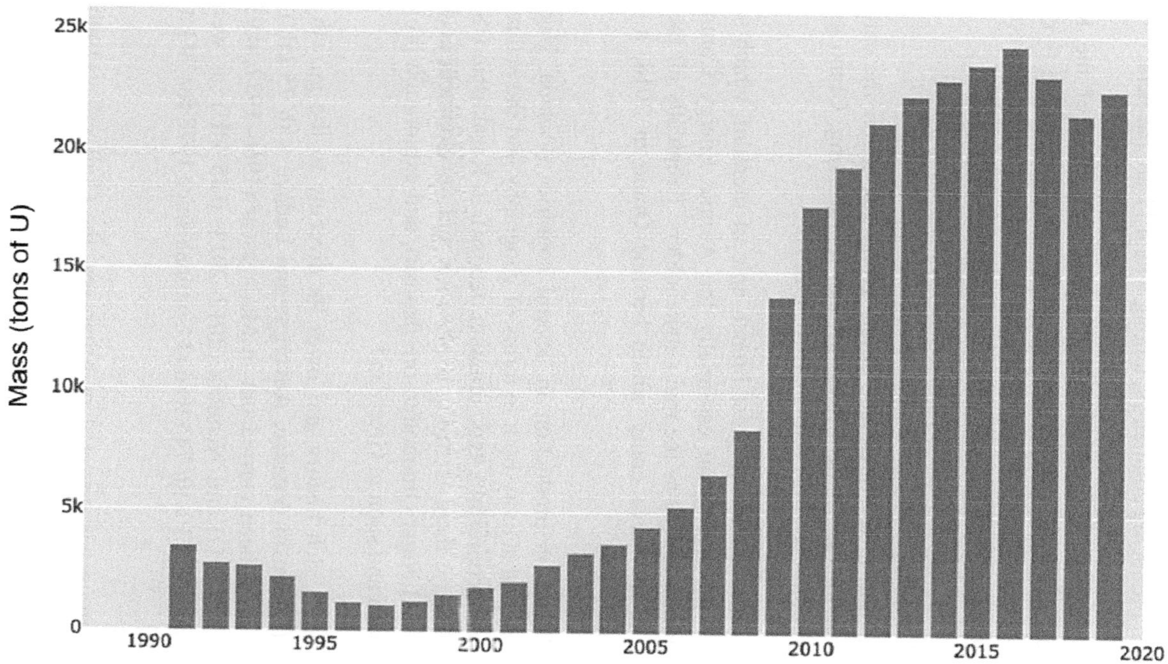

Figure 4.13. *Production history of Kazakhstan (source: IAEA). For a color version of this figure, see www.iste.co.uk/percebois/nuclear1.zip*

4.2.5.2.1. Kazatomprom

The state-owned company Kazatomprom, which recently opened up its capital (eventually 25% on the market), is the world's leading producer of uranium. It is present in 15 uranium producing companies, among which 11 are partners.

Kazatomprom[3] controls 20% of the world production, 60% of the Kazakh production, produced solely by ISR technology in the Shu-Syrasu and Syr Darya basins. Its strategy is based on very low production costs and abundant resources. Kazakhstan's production growth has been meteoric (Figure 4.13).

4.2.5.2.2. Orano Mining

Orano Mining[4] is part of the integrated Orano group that covers the entire fuel cycle: mining, conversion, enrichment, logistics and reprocessing. Orano group is derived from the restructuring of the Areva group in 2017.

Orano Mining is the heir to the great French epic of uranium, which saw the appearance of many producers, but essentially the Atomic Energy Commission (CEA) and then Cogema (1976) as the main stakeholders. Eventually all the assets were concentrated in an Areva entity which became Orano in 2018, a French state-owned company.

Its strategy is based on production in several countries (Canada, Niger and Kazakhstan), a diversified activity both in terms of geography and the nature of the mines: in Athabasca (unconformity-type uranium deposit), in Kazakhstan (ISR) and Niger (sandstone). Today, there is strong exploration and development activity in Mongolia, Namibia and Niger (Imouraren ore deposit), and exploration projects in Uzbekistan and Canada and soon in Greenland.

The company has a rich history and owes its diversity to the growth of the French production (Limousin, Vendée, Auvergne and Lodève with 75 ktU). Other production projects have been carried out in Madagascar (1966–1977, 0.95 ktU), Gabon (1961–1998, 25 ktU) and Namibia (2012–2013, 0.45 ktU), as well as in Niger with Somair since 1971 with 74 ktU and Cominak since 1978 for 79 ktU, until its shutdown in 2021.

In Canada, the group has operated in Athabasca, the Cluff mine (1980–2003, 22 ktU, 100%) and MacLean (1999–2010, 19 ktU, 70%) where the treatment plant

3 Available at: https://www.kazatomprom.kz/en.

4 Available at: https://www.orano.group/fr.

has worked with ores from the Cigar Lake mine since 2014 (25 ktU cumulative production; 37.1% Orano interest, operated by Cameco).

The other participations in Cameco mines have totaled 16.67% in the Key Lake mine (1983–2001, 80 ktU) and 30.2% in the McArthur mine (125 ktU produced since 2000).

Orano Mining controls 51% of Katco (49% Kazatomprom) in Kazakhstan, where ISR production began in 2006 and reached 42 ktU by the end of 2019. The operations were set on two sites, Muyunkum and Tortkuduk.

Taking into account McArthur's production (under maintenance in 2018) and considering all the Nigerian production under its control for the two mines operated by Orano, this firm is considered the second global stakeholder.

4.2.5.2.3. ARMZ – Uranium One (U1)

This Russian state-owned company from the Rosatom group (which covers the entire nuclear cycle and owns 38 reactors in Russia) is divided into two entities, ARMZ[5] (Atom Redmet Zoloto) and Uranium One (U1).

ARMZ is in charge of operating three Russian mines, including the oldest uranium mining complex worldwide, in the Strelstovskoye's caldera (Eastern Siberia). Built in 1970, several underground mines have produced nearly 150 ktU, and in 2019 it was the complex producing the most uranium in the world. It is operated by the subsidiary company JSC Priargunski.

The other two producing areas are the ISR fields of Khiagda near Lake Baikal and Dalur in the southern Urals.

Uranium One[6] (U1) operates mines outside Russia and produces between 4,500 and 5,000 tU/year via six joint ventures with Kazatomprom in Kazakhstan, where it started its production in 2007.

The small Willow Creek mine in Wyoming produced 1,000 tU between 2011 and 2018, and it is currently undergoing maintenance.

U1 also holds the Mkuju sandstone deposit in Tanzania, which has not been profitable to date (45 ktU, with 300 ppm U_3O_8).

5 Available at: https://rosatom.ru/en/rosatom-group/uranium-mining/.

6 Available at: https://uranium1.com/about-us/.

4.2.5.2.4. Cameco

This Canadian[7] firm listed on the Toronto Stock Exchange, displays a strong mining activity at Saskatchewan (Canada), where it operates the McArthur mines (under maintenance since 2018), Cigar Lake (in operation, but temporarily suspended due to COVID-19 between March and September 2020) and Rabbit Lake (under maintenance since 2016).

Outside Canada, in Kazakhstan, Cameco has a joint venture with Kazatomprom at Inkai, with a 40% stake (60% before 2018), nearly retaining 15 ktU.

In Wyoming, Cameco operates two ISR mines, Crow Butte and Smith Ranch which have produced approximately 17 ktU since 1991. These have been under maintenance since 2018.

Cameco also holds interests in numerous deposits and exploration areas in Athabasca.

4.2.5.2.5. Diversified miners: BHP and Rio Tinto

Being the second and third global mining companies, BHP and Rio Tinto produce uranium, although this represents less than 1% of their turnover. Even if the role of BHP is potentially growing and Rio Tinto could disappear, these two are remarkable stakeholders.

BHP owns the world's largest Olympic Dam Deposit (ODM[8]) in South Australia, but this is primarily an important underground copper mine. Uranium is not BHP's main motivation, but rather its consequence. Any program destined to increasing this mine's capacity first has an impact on copper production and therefore on uranium!

BHP wants to increase its annual copper capacity from 180 to 300 ktCu by 2025, proportionally from 3,200 to 6,500 tU, which is why the consequences on the market are not negligible. However, the road will be long because it is necessary to solve serious technical problems regarding the drilling of galleries and winzes, and treatment difficulties, with a heap-leaching passage plan. Will the mine's efficiency meet the expectations? This is a matter of great concern for the uranium market.

Rio Tinto had two uranium mines, one in Namibia, Rössing, which was sold to the Chinese company CNNC in 2019, and the Ranger in the Northern Territory in

7 Available at: https://www.cameco.com/.

8 Available at: https://www.bhp.com/our-businesses/minerals-australia/olympic-dam/.

Australia, still belonging to Rio Tinto and operated by Energy Resources Australia (ERA[9]).

The Ranger mine ceased its activity at the beginning of 2021, having been unable to obtain the support from the aborigines to continue mining underground on Ranger Deep, despite the construction of a winze.

4.2.5.2.6. The Chinese mining industry

China's procurement policy is based on a three-pillar premise:

– one-third drawn from domestic resources;

– one-third purchased on the international market;

– one-third of mines outside China, but operated by Chinese companies.

To date, two Chinese companies have uranium mining activity, but Chinese private funds, particularly from Hong Kong, are acquiring stakes in exploration companies, particularly in Australia and Canada.

The state-owned group CNNC (China National Nuclear Corporation[10]), controls all the uranium production in China, covers the fuel cycle and produces electricity (21 reactors in service and six under construction). The group is increasingly oriented toward ISR technology in China, with mines in Xinjiang and Inner Mongolia, which will eventually supply all the production.

However, ISR technology will have to rely on alkaline solutions due to the presence of limestone in the formations. This could have an impact on the cost, since alkaline solutions are more expensive than acid injections.

The Fuzhou underground mine in the south, probably the deepest in the world (but rather poor), will probably shut down in this decade.

CNNC has invested outside China, in Niger, where it opened the Azélique mine between 2011 and 2013 when the uranium prices were at their highest. At the moment, 666 tU were extracted but since then the mine has been under maintenance.

In 2014, CNNC took a 25% stake in Paladin's Langer Heinrich mine in Namibia, although the Chinese could not benefit for long, because the mine went into maintenance in 2018.

---

9 Available at: https://www.energyres.com.au/.

10 Available at: http://en.cnnc.com.cn/.

Also in Namibia, CNNC bought out Rio Tinto in 2019 and operates the Rössing[11] mine. Despite being the poorest uranium mine in the world, the resources are still important but the ore is changing (SJ area), and the more expensive treatment will require new facilities. A recovery project involving heap-leaching of low-grade ore (<250 ppm), which has already been mined could be relaunched; in that sense, Rio Tinto has envisaged 16 years of additional extraction.

CNNC controls another deposit in Namibia, Anomaly 18, which is poorer than Rössing. In Mongolia, in 2009, it invested in the developing ore deposit of Gurvanbulag, whose resources are modest.

CGN (China Guandong Nuclear[12]) is a powerful nuclear utility with 16 reactors in service and 11 under construction. It began to show its interest in uranium in 2009 by taking control (66.45%) of a junior mining company, Energy Metals, who holds an export license for a small deposit in Australia, but whose activity is under maintenance.

The big investment for CGN was the acquisition of the Husab deposit for 2.2 billion US$ in 2012, operated by Swakop Uranium[13]. This Alaskan-type deposit is close to that of Rössing, but with a slightly higher grade. Production started in 2017 and in 2019, and it produced 4,240 tU. CGN hopes to reach the nominal capacity of 5,770 tU in 2021–2022. All the production is destined to China.

4.2.5.2.7. Other actors

Among the remaining 15%, we can mention a few marginal producers who do not take part in the international market and keep their production (Iran, India, Pakistan, Brazil, Ukraine). Despite their admirable development efforts, ore deposits are low-grade and expensive.

Only a few producers can (or desire to) produce at an international scale.

Navoi, the Uzbek state-owned company, does not communicate much due to legal and political reasons inherent in their country, but estimates show that their production is substantial, around 2,500–3,000 tU/year, with high export levels. They produce exclusively with SRI technology, which accounts for the relatively low costs.

---

11 Available at: https://www.rossing.com/.

12 Available at: http://en.cgnpc.com.cn/.

13 Available at: http://www.swakopuranium.com/.

The Australians from Quasar (subsidiary of the American firm General Atomics) produce 2,000 tU/year, in South Australia at the Beverley mine.

As already mentioned, the production in South Africa is closely related to the price of gold (uranium being an accompanying by-product). At present, Harmony Gold produces around 350 tU/year.

Although the family of American ISR producers (Ur-Energy, Peninsula, EFI, UEC) yields a small production (200–800 tU/year), this is highly sensitive to the market price, which means it can quickly start and stop.

Three mines have been shut down in recent years, their cost making it impossible to continue: Langer Heinrich (Paladin), Kayelekera (Lotus) and Honeymoon (Boss). Depending on the price, they can return to the market in less than 2 years, with an initial production of 1,200 tU/year for Paladin, probably 1,000 tU/year for Lotus and 700 tU/year for Boss.

4.2.5.3. *Term and spot contracts*

Uranium has the particularity of having only one use: the nuclear reactor. The end customer is therefore the electricity supplier. The market has two quotes available for prices: the spot price and the long-term price, as explained in Table 4.3.

	Spot Delivery between 3 and 18 months	Long-term Multi-year contract with delivery > N + 3
Annual volume (2019)	40% (~22kt U)	60% (~32 ktU)
Average transaction	50–100t ~3-6 10^6 US$	250 t/year over 5 years ~100 10^6 US$
Main customers	Electricity suppliers, investment funds, producers, traders	Electricity suppliers
Purchase criteria	Financial opportunities: stock management, speculation	Security, reliability, diversification of supply sources (countries, suppliers), price visibility
Price mechanism	Fixed price	Basic price (predictability) Market price (on delivery) Mix
Indicators (2019)	US$25/lb$U_3O_8$	US$31/lb$U_3O_8$

Table 4.3. *Characteristics of the uranium market quotes (source: Orano)*

4.2.5.4. *Uranium price*

Uranium prices have greatly varied over the past 50 years, not only under the influence of the nuclear power program (1970s decade), but also oil events (1973), political events (Megatons to Megawatts USA/USSR Agreement 1993–2013), nuclear accidents (Three Mile Island 1976; Chernobyl 1986; Fukushima 2011) and economic crises (2008) (Figure 4.14).

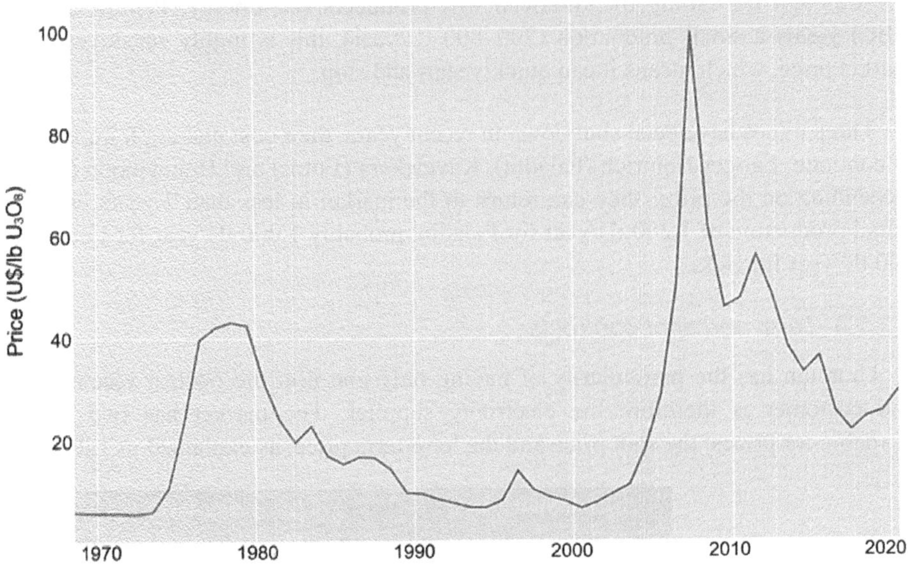

Figure 4.14. *Average uranium spot price. For a color version of this figure, see www.iste.co.uk/percebois/nuclear1.zip*

Regarding the future of uranium costs, it is worth noting that any mine can produce as long as the market absorbs its production. In a competitive market, uranium mines will be classified into three main categories:

– ISR mines have the lowest costs; they can be estimated between 15 and 25 US$/lbs.

– The Athabasca mines, with high grades but expensive technologies (jet boring), have costs oscillating between 20 and 30 US$/lb.

– The other mines have costs exceeding 25 US$/lbs.

On announcing these costs, the recent price period put many mines in difficulty, in particular the mines of Langer Heinrich (Namibia), Kayelekera (Malawi),

Honeymoon (Australia), the ISR mines in the United States (in an alkaline medium), as well as projects such as Imouraren (Orano/Niger) whose onset has been delayed.

One may separately consider the mines whose costs are potentially high but whose production is maintained, either due to strategic (Husab/CNNC in Namibia) or to political reasons (India, Iran, China, Brazil and Pakistan). However, it should be noted that in the latter case, their production is marginal for the market.

When uranium is a co- or by-product, the mining costs are borne by the main metal production, and therefore, the cost of extracting uranium is low. This is the case for the Olympic Dam or South Africa, for which the production of uranium is maintained, in addition to copper or gold production.

Temporary shutdowns and recent maintenance such as McArthur or Rabbit Lake correspond to temporary marketing decisions, given the low market prices and large stocks. It is preferable to preserve low-cost capacities until the prices improve.

We have already seen that the increase in extraction costs prompted the IAEA to create a new higher cost tranche, as well as an increase in the average costs of the resources to be mined (Table 4.1). This is true for uranium as well as for the mining industry in general. According to McKinsey (Table 4.4), the operational and investment costs sharply increased between 2004 and 2013, and since 2014, mines have had to make enormous efforts to stabilize or reduce their costs, even to the point of shutdown.

Annual growth	Operational costs	Investment costs
2004–2009	14.5%/year	11.5%/year
2009–2013	7.5%/year	10.6%/year
2014–2019	0.2%/year	–0.5%/year

Table 4.4. *Increase in operational and investment costs in the mining industry (source: PDAC Toronto 2020)*

4.2.5.5. *Supply security*

Substantial investments in the nuclear sector must be guaranteed by reliable and long-term uranium supplies.

As has been shown for uranium resources, despite a cumulative production of nearly 3 MtU since 1945, the resources have continued to grow, together with their quality.

Reactor lifespan has also continued to increase and could reach 80 years. As a result, the electricity supplier needs to develop a long-term perspective. Uranium producers have the responsibility of ensuring this supply, as well as diverse choices, due to strategic and political reasons. National independence in the energy field is a key subject in the politics of countries.

The last 15 years have shown that, if necessary, the mining industry can quickly invest to bring in new resources. However, the characteristics of ore deposits change. This is a general movement affecting the entire mining industry: deeper, poorer and more distant ore deposits.

Despite the fact that resource exploitation is more expensive (Table 4.1), the price of uranium always remains marginal in relation to the cost per kW/h produced.

The question of industry independence in the face of supplies coming from another country – with the accompanying systemic risks – is essential.

Three interpretations or risk levels are possible:

– the case of a foreign supplier, in which case the fabrication facility is dependent, at risk;

– the case of a fabrication facility owning and controlling a mine, in which case the risk is mitigated;

– the fabrication facility owning and controlling several mines in different countries, in which case the risk completely disappears.

It is this last point that Orano and its predecessors have developed for uranium: controlling mines in various countries to supply the French demand.

4.2.6. *Prospecting and exploration expenditures*

Knowledge of uranium geology and prospecting have made great progress since the end of the II World War. It was only after the war and during the atomic era that uranium geology took off. Marcel Roubault, Director of the School of Geology in Nancy and Director of the CEA after the war, gave the impetus for systematic and thoughtful prospecting, with an ambitious plan sending many geologists to the French countryside with a Geiger counter in hand.

More than 900 million US$ were invested by the French mining industry for the exploration of uranium until the end of the French production (2002), against 13 billion US$ destined to worldwide exploration during the same period.

4.2.6.1. *Development of a mining project*

A mining project is the succession of different stages which can be continuous, but also interrupted at each stage, depending on the market (Figure 4.15).

4.2.6.2. *Exploration*

At the beginning, exploration is launched based on geological and historical considerations (prior mining activity, historical data and prospecting). It often starts with global aerial coverage measuring conductivity, magnetism and radioactivity, information which is plotted onto geological maps. Targets are confirmed by field measurements and prospecting, and sample collection work (geochemistry). If the data reveal the discovery of an anomaly or evidence, then reconnaissance surveys are carried out. The drilling mesh is then narrowed in order to obtain a better image of the anomaly, which could eventually become an ore deposit due to resource identification (and hopefully of reserves, depending on the results of successive studies). In-depth studies include ore knowledge (mineralogy), the geometry of the ore deposit, treatment metallurgy, geotechnics, hydrology and the elements relating to the environment and the communities involved. Environmental and societal studies are fundamental to defining the initial conditions of the zone, information which will become crucial during refurbishment works. Everything will depend on the existing initial information, market pressure, the ease of access, acceptance by the population, all of which can take many years before start.

4.2.6.3. *Exploitation*

Launching the construction of the mine and the plant is highly dependent on the authorization deadlines. Several years are often necessary.

Take the Athabasca ore deposits, for example. Figure 4.16 shows the difference between the deposit's discovery and exploitation.

For the uranium mine, specific extraction technologies have been developed, such as ISR which is also applied to high-grade deposits. In Canada, the aquifer can be very present there, involving subsoil freezing to act as a barrier, coupled with extraction by hydraulic blasting (jet boring). Needless to say, this process is fully automated so that no miner comes into contact with the ore.

Throughout their lifetime, some mines come to a halt for a few months or a few years and resume their activity, depending on economic circumstances.

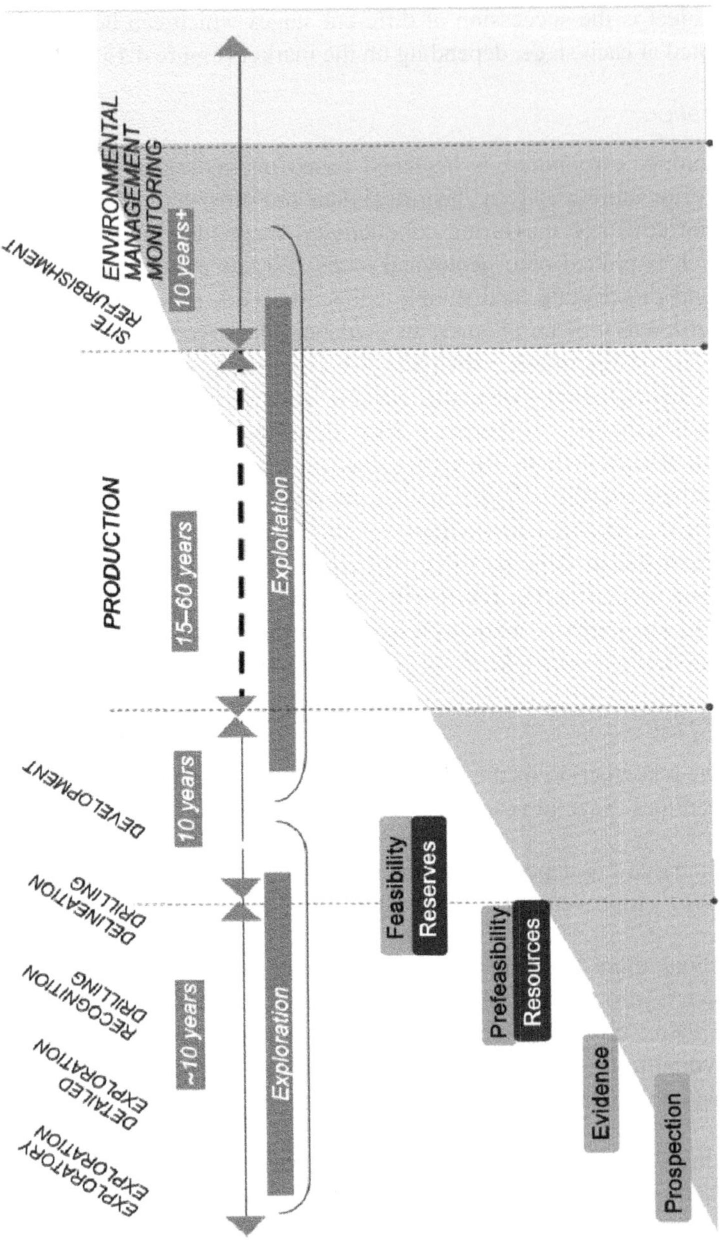

Figure 4.15. *Stages of a mining project (source: Orano). For a color version of this figure, see www.iste.co.uk/percebois/nuclear1.zip*

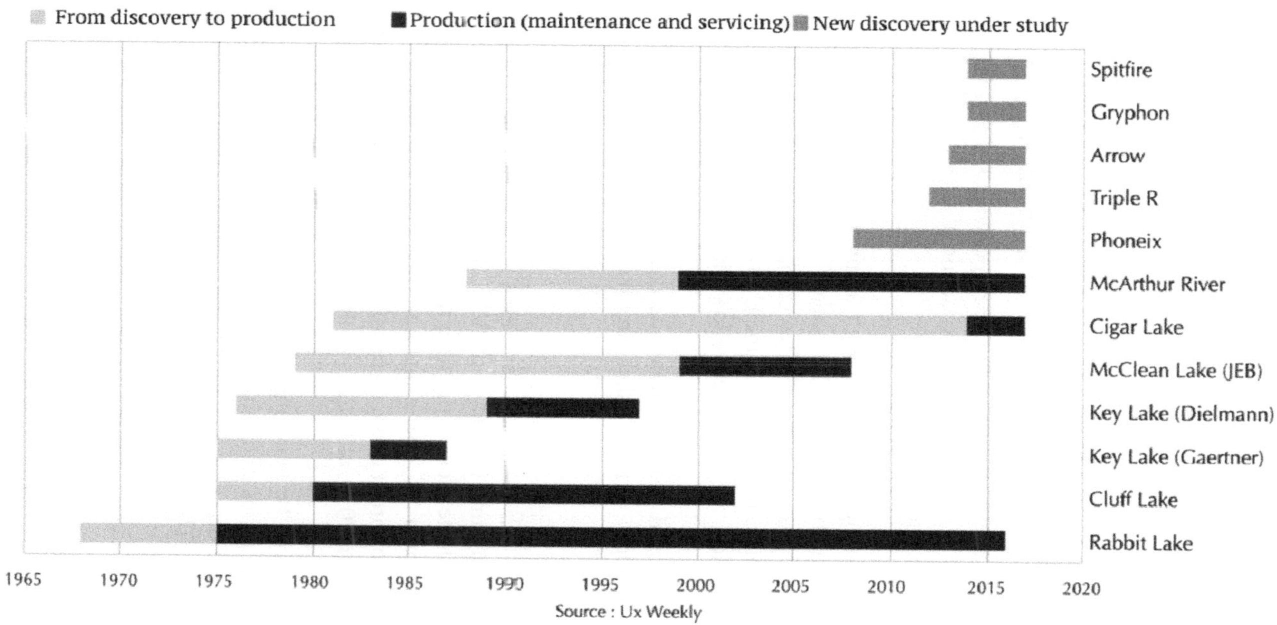

Figure 4.16. *Mining development schedule for the Athabasca ore deposits (source: Ux Weekly 1995). For a color version of this figure, see www.iste.co.uk/percebois/nuclear1.zip*

The ore deposit's richness, extension and prices will determine its longevity. But the mine's policy is clearly to persist! The oldest mines (in order of permanence) are Strelstovskoye (1970), Somaïr (1971), Rössing (1976), Cominak (1978), Ranger (1981), the Olympic Dam (1988), McArthur (2000) and Cigar Lake (2014), which means that the mine settles permanently when it can.

The treatment schedule for uranium ore is relatively basic: the ore undergoes physical preparation (crushing/grinding) before being sent to acid dissolving tanks (sometimes alkaline solutions).

Once the solution loaded with uranium has been generated, the ordinary process is to open pit, underground or ISR mines: the extraction of uranium is done either by resins or by liquid/liquid extraction.

The purified solution is highly enriched with uranium, precipitated and dried to form concentrates traditionally called "yellow cakes" because of their color, which are the plant's marketable products. The yellow compound is either sodium (SDU) or ammonium (ADU) uranate, but today it is the U_3O_8 or sometimes the OU_4, which are privileged. Although these are no longer yellow, the term has remained.

Thus, the ore is processed near the mine and does not travel, contrary to what was done in the past. Only yellow cakes are transported and exported to the converter.

4.2.6.4. *Refurbishment*

The decision to close a mine is always difficult; the most obvious reason is the exhaustion of reserves and the lack of possible extension. Refurbishment generally begins before the shutdown of operations on the mine's depleted areas, and sometimes before the end of production, once stocks have built up. The Ranger mine ceased its mining activity in 2013, although its plant produced uranium until 2021.

Refurbishment works can be very important, and their costs around 1–2 US$/lb produced. The studies, analyses and environmental monitoring will be carried out over several years after the cease of operations; the residual production, in particular of mine water, is possible. The Lodève mine, closed in 1999, continued to produce 1–2 tU/year, 20 years later.

4.2.6.5. *Prospecting–price correlation*

Mining exploration, financed by Junior Mining fundraising or the dedicated budgets of mining companies, responds to market demands. The correlation between commodity prices and funds allocated over the past 15 years for uranium is strong, but with a 1-year lagged response (Figure 4.17).

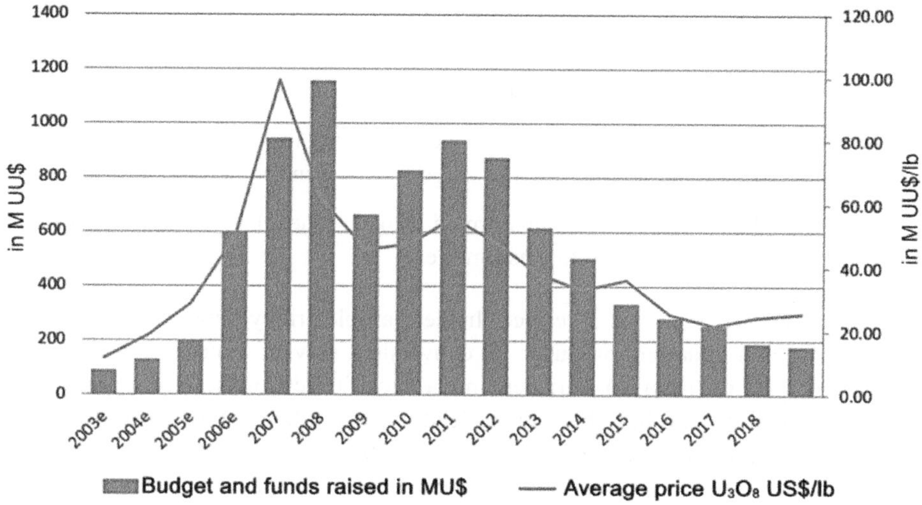

Figure 4.17. *Amount of funds raised for uranium exploration (source: S&P, Orano). For a color version of this figure, see www.iste.co.uk/percebois/nuclear1.zip*

Over the period 2005–2011, although a nuclear power revival was expected worldwide, the profound impact of the 2008 crisis as well as the consequences of the Fukushima accident were only reinforced.

It should be noted that these financial efforts are considerable and have not gone unnoticed for the new resources identified. Between 2005 and 2018, new 1.3 MtU resources were discovered.

4.2.6.6. *To conclude this section*

The uranium industry has demonstrated its resilience and its ability to adapt to the needs of the nuclear industry. International actors, be them state-owned or private companies, make this metal available for costs are always economically compatible with the industry's development.

4.3. From conversion to fuel fabrication

In section 4.2, we reviewed the uranium resource. This uranium from mines or secondary resources is generally available under the chemical form U_3O_8. Under that chemical form, uranium cannot be used for making fuel for nuclear power plants. Various additional stages are required before being able to use it. These are listed below and are the subject of the following sections:

– the first stage involves changing uranium's chemical form, transforming it into UF_6. This is a fluorination stage called conversion;

– UF_6 can then be enriched, which means increasing the percentage of the uranium-235 isotope in the uranium; this stage is called enrichment;

– enriched UF_6 can then be transformed into enriched UO_2 which will be the basic material used to fabricate fuel assemblies.

These stages are sold as a service. Indeed, an electricity supplier buys natural uranium from a uranium producer, a conversion service from a converter, an enrichment service from an enricher and a fuel fabrication service from a fuel fabrication facility. The first three stages lead to standardized products so that these stages are completely independent and can be subcontracted to different suppliers, which results in a highly competitive market: actually, it is a commodity market. It should also be noted that very few companies are present throughout the cycle: many companies specialize in a single activity.

4.3.1. *Conversion*

This is the first stage in the uranium transformation cycle, whose goal is to transform the yellow cake (usually from the U_3O_8) derived from the UF_6 mining industry. This stage is also accompanied by a purification stage. The purpose of this chemical transformation is to obtain a chemical form of uranium enabling its enrichment. Conversion plants are also the place where yellow cakes are stored, for different reasons: strategic stocks, commercial stocks, etc.

4.3.1.1. *Processing at the conversion plants*

Generally, an electricity supplier buys a conversion service: the supplier has the uranium mining concentrates (the yellow cakes purchased from the mining industry), directly delivered to the conversion plant. These concentrates are packaged in drums at the mines where they are produced and then transported by sea or railway in ISO containers, meeting specific and restrictive transport standards. These drums are then unloaded and stored at the plant and serve as the plant's feed material. The uranium is then converted into UF_6 and stored in a specially designed packaging for this purpose: 48Y. These are metal cylinders containing up to 8.5 tU. These packages are filled with gaseous or liquid UF_6, after which UF_6 is solidified. It should be noted that this UF_6 presents chemical risks in contact with air, because it is transformed into hydrogen fluoride (HF) via a chemical reaction with the water vapor contained in the air.

Given the fungible and undifferentiated nature of the material received, the atoms constituting the uranium sent by the electricity supplier A, are not necessarily those which will constitute the UF_6 assigned to this same electricity supplier A. They may have been interchanged with those sent by electricity supplier B or C. At the end of the transformation, the UF_6 is analyzed and must satisfy a single standard defining all the necessary parameters: the ASTM C787 standard.

4.3.1.2. *Material codes*

A code system is applied to uranium and its conversion, reflecting the countries where the uranium was mined and where it was converted. These codes were put in place to control the use of uranium, which can be a dual-use resource and reflect the agreements between the different countries around the world. As a matter of fact, some countries refuse to allow uranium to be used for purposes other than civilian ones, or simply by certain third countries. This can be a constraint for converters.

4.3.1.3. *The conversion market*

4.3.1.3.1. General view

During the creation of assemblies, despite being an unavoidable stage, conversion has always represented the smallest part of the value chain.

Over the past 10 years, production costs have increased because conversion plants have had to deal with new constraints imposed by the countries' nuclear authorities after the Fukushima accident. On the other hand, during this same period, demand fell following some government decisions to phase out nuclear power, or to postpone the construction of new nuclear power plants for the same reasons: public opinion acceptance after the Fukushima accident. Finally, the reduction in the strategic stocks of nuclear weapons has fueled the market for enriched uranium without going through the traditional process (uranium from mines, conversion, enrichment). These three phenomena, combined with the continued operation of all of the world's production capacity, led to overcapacity on the market and increased stocks. The market price of the conversion therefore fell.

In recent years, however, measures have been taken by converters. Note, for example, the capacity reduction in Cameco (Canada) or Rosatom (Russia), the suspension of activities at Converdyn (United States), or the permanent shutdown of Springfields (England).

All these measures are now making the world conversion production capacity much lower than the annual demand: while the market has started to rise again, the stocks created after Fukushima have decreased. This is illustrated in Figure 4.18.

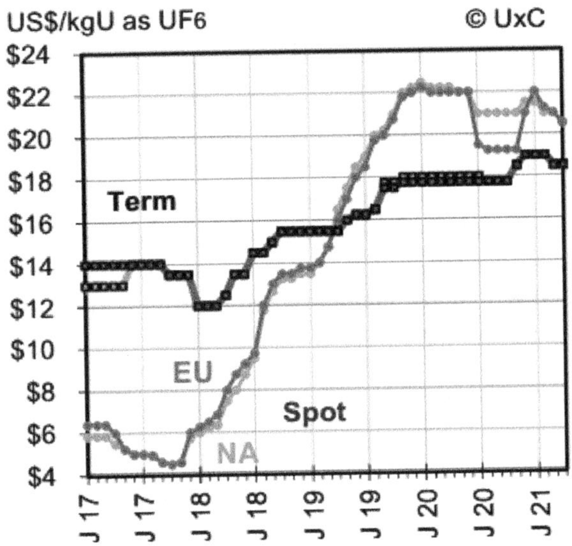

Figure 4.18. *Evolution of the conversion price from 2017 to 2021 (source: UxC Market Outlook, Update 2021, UxC). For a color version of this figure, see www.iste.co.uk/percebois/nuclear1.zip*

4.3.1.3.2. Demand forecasting

The new projections for the conversion demand of the next 20 years predict that at the very least, the demand will stabilize, as shown in Figure 4.19.

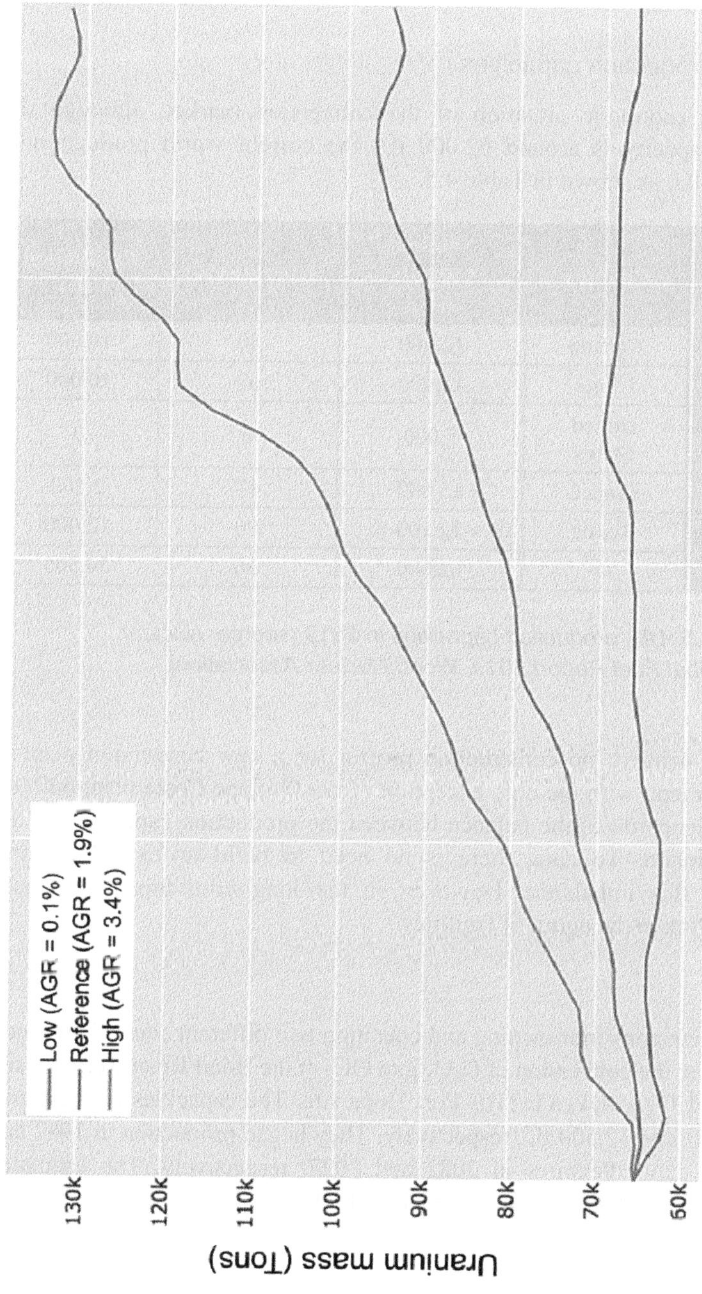

Figure 4.19. *Conversion demand scenarios until 2040. AGR stands for annual growth rate (source: Nuclear Global Fuel Report 2019, World Nuclear Association). For a color version of this figure, see www.iste.co.uk/percebois/nuclear1.zip*

This is to be compared with the drastic reductions in production capacities carried out by converters.

4.3.1.3.3. Global production capacities

Considering the economic situation of the conversion market, although the nominal installed capacity is around 62,000 tU, the current world production is estimated at 34,500 tU, as shown in Table 4.5.

Converter	Country	Nominal capacity (tU)	Use rate (%)	Production (tU)
Cameco	Canada	12,500	80	10,000
CNNC	China	15,000	67	10,000
ConvertDyn	United States	7,000	0	0
Orano[14]	France	15,000	17	2,500
Rosatom	Russia	12,500	96	12,000
Total	–	62,000	56	34,500

Table 4.5. *UF_6 production capacities in 2019 (source: Nuclear Global Fuel Report 2019, World Nuclear Association)*

In this alamort context, no construction project for a new conversion plant is planned for the moment, with the only exception of the Philippe Coste plant built by Orano. This issue jeopardizes the balance between the production capacity and the long-term UF_6 demand. To date, there is no need to build or extend existing capacities, despite this imbalance. However, in the long-term, this point could become necessary due to the aging of facilities.

Cameco

This is a Canadian converter owning and operating two different sites. Conversion is done in two stages: the conversion of U_3O_8 into OU_3 at the Blind River Refinery site and the shift from UO_3 to UF_6AT THE Port Hope site. The capacities of these two plants are 24,000 tU and 12,500 tU, respectively. They began production in 1982 and their authorization expired/expires in 2022 and 2027, respectively. The estimated quantities of UF_6 produced in 2018 were around 10,000 tU.

14 The conversion capacity utilization rate of Orano is increasing and should probably reach 100% in the next few years.

Converdyn

This is a joint venture between Honeywell and General Atomics, which has operated the Metropolis plant in the United States since 1982. The nominal capacity of this plant is 15,000 tU. The conversion is done in one stage.

In order to improve the economic conditions of the conversion market, Converdyn first decided to significantly reduce its annual production to 7,000 tU in 2017, before completely shutting down its production a year later because the market price had fallen below its production costs. In this context, Converdyn has concluded agreements with various converters to ensure their own sales. On the one hand, this led to a substantial reduction in stocks and, on the other hand, to an increase in market prices.

At the same time, Coverdyn has invested in maintenance for its plant, so that it can quickly resume activities once all purchase contracts have been settled. It should be noted that, during the plant's shutdown, Converdyn maintained the storage, sampling and weighing of materials on its site.

Orano

Orano, formerly Areva, is an industrial group whose activities include the entire fuel cycle: mining, conversion, enrichment, logistics and reprocessing, as well as decommissioning and associated services. The group is derived from the re-structuring of Areva in 2017, having refocused on cycle activities.

Orano has been operating a conversion plant called Comurhex for almost 60 years, located on two different sites: the Malvesi plant (which turns yellow cakes into UF_4) and the Tricastin plant (which transforms UF_4 into UF_6).

In 2018, Orano inaugurated a new UF_6 conversion plant on the same sites called Philippe Coste, the name of the creator of the previous Comurhex plant. The Comurhex plant closed at the end of 2017, after having produced more than 450,000 tU during its years of operation. Many technical innovations have been integrated into this plant, which has the particularity of accepting all types of mining concentrates.

Rosatom

This is a Russian state-owned company, present in all the cycle activities and associated services. For the most part, Rosatom does not sell conversion on the market directly as a service, but rather integrates it into the sale of fuel assemblies.

In order to adapt to the market after Fukushima, Rosatom closed two conversion plants in 2014 and 2016, at its Angarsk and Glazov sites. Conversion is now taking place at the Seversk site and numerous investments have been made to modernize it and make it more reliable. Plans to create new conversion plants have been postponed until market economic conditions notoriously improve.

CNNC

China has its own conversion plants whose goal is multiple: not only to supply the Chinese domestic market, but also to be the supplier of Chinese reactors built abroad and to find new customers on the European and American markets.

Currently, CNNC has the capacity to supply with UF_6 the entire fleet of operational reactors located on Chinese territory. A new phase involving an increased production capacity is planned, but the information relating to this subject has not been disclosed.

4.3.1.3.4. Market share of the respective stakeholders

Figure 4.20 shows the distribution of the production forecast for the years 2019–2023 where four major stakeholders share the production in a balanced way. The forecast for primary production is 213,000 tU for an overall capacity of 69%.

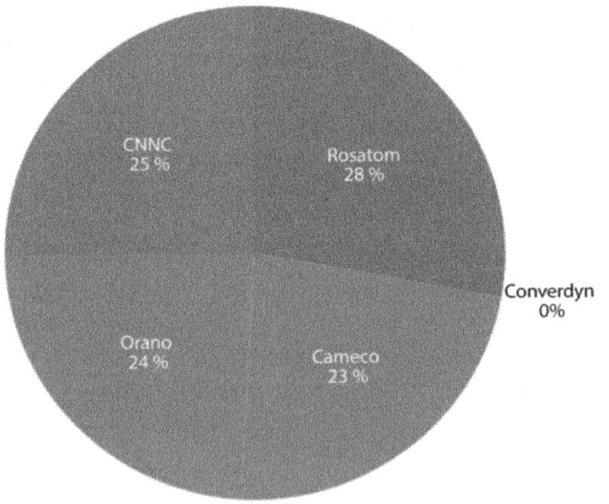

Figure 4.20. *Production forecast by stakeholder for the period 2019–2023 (source: Nuclear Global Fuel Report 2019, World Nuclear Association). For a color version of this figure, see www.iste.co.uk/percebois/nuclear1.zip*

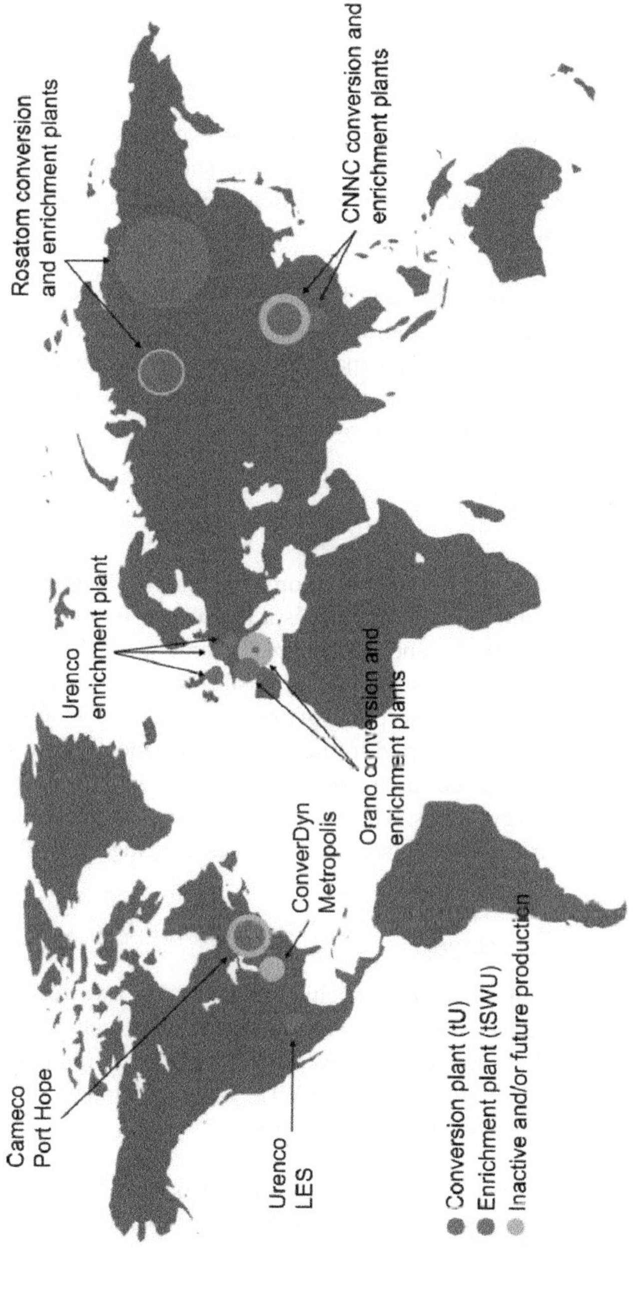

Figure 4.21. *Location of UF_6 production and demand from enrichers in 2019. For a color version of this figure, see www.iste.co.uk/percebois/nuclear1.zip*

As Figure 4.21 illustrates, European production does not fully satisfy the demand at present. This balance should be reached when the Philippe Coste plant in Orano has attained its nominal production.

4.3.1.4. *Conclusion*

Despite a particular economic situation intimately related to the Fukushima accident, the scenarios for the years 2020–2040 either show a stabilized conversion demand in the case of the lowest scenario or a strong (or even very strong) increase in the medium and high scenarios.

In relation to the short-term capacities, although insufficient, this deficit could be filled by the stocks created during the previous decade. In the medium-term, it is undoubtedly possible to see the frozen capacities restart, depending on the market's economic conditions. Finally, in the long-term, new capacities could be required, either by increasing existing capacities or by creating new plants.

4.3.2. *Enrichment*

This is the second stage in the transformation of uranium. The goal of this stage is to enrich natural uranium. Natural uranium is a mixture of uranium isotopes, the two main ones being ^{238}U (99.3%) and ^{235}U (0.7%). Uranium-238 is a stable isotope and uranium-235 has the characteristic of being fissile, which means it can be split into several fragments after the absorption of one neutron. The kinetic energy of these fragments can then be transferred to the medium and locally heat it. It is this property that is sought during the thermonuclear reactions, which take place in the reactors from nuclear energy plants. The enrichment operation involves increasing the proportion of the uranium-235 isotope to values ranging from 2 to 5%, as needed.

4.3.2.1. *Enrichment technologies*

There are several technologies for enriching uranium, but two technologies are the most commonly used in the fuel cycle industry. One technology produces enriched uranium and the other technology produces depleted uranium. The unit measuring the separation work is called SWU. Any enrichment work can be converted to SWU, regardless of the technique used.

The first technology used in the past but which has now become obsolete is enrichment by gaseous diffusion. The first French enrichment plant (Eurodif), which shut down in 2011, used gaseous diffusion. This technology passes uranium in the form of UF_6 gas through different compression stages and filters (called diffusers), creating two different flows: one of depleted uranium and the other of enriched uranium. At each passage through a diffuser, the enriched uranium stream is slightly enriched. The same happens with the depleted stream, which becomes increasingly depleted. The advantage of this technology is its flexibility of use and the fact of being able to easily produce any enrichment grade, between 1% and 5%. The weak point of this technology is its electrical cost since the many diffusers and compressors (more than 1,400 at Eurodif) consume the production of three nuclear reactors when producing at full speed.

The second technology, the most widely used in our days, is enrichment by centrifugation. It relies on isotopic separation, but exploiting the difference in mass between the isotopes. The UF_6 gas flow passes through a first centrifuge, creating two different flows (an enriched flow and a depleted flow). Then each stream passes through a new centrifuge, further enriching the enriched stream and making the depleted stream poorer. After passing through a large number of centrifuges, the enriched uranium can reach the values required for manufacturing the fuel assemblies used in electricity-generating nuclear power plants. The main advantage of this technology is that it reduces electricity needs by a factor of 50, compared to gaseous diffusion technology.

There is a third technology for enriching uranium, based on the use of lasers. However, to date, this technology is still under research and needs further development. It has not yet been industrialized.

4.3.2.2. *The operation of enrichment plants*

As for the conversion, an electricity supplier buys an enrichment service: the delivery of natural UF_6, in 48Y cylinders, generally purchased from a converter. This uranium is transported in 48Y containers by sea or land and deposited at the bottom of enrichment plants. The packages are introduced into the plant, heated in order to sublimate the solid UF_6 into UF_6 gas, and fed to the centrifuges.

After enrichment set at the desired level (called grade or enrichment rate), the enriched uranium is stored in a new type of container: 30B. The 30B is a smaller cylinder than the 48Y and can hold up to 1.5 tU in solid form. Depleted uranium is stored in a 48Y. The percentage of uranium-235 in the depleted stream is set according to market conditions and called "tails assays". It should be noted that ownership of depleted uranium amounts to the enricher, who manages and stores it. Before storage, it is usually deconverted into U_3O_8, which is a chemically stable and less hazardous form than UF_6. It is worth noting that this uranium is considered a material (not a waste), because it can be re-enriched.

As for natural UF_6, enriched uranium is fungible and undifferentiated. Therefore, the atoms making up UF_6 sent by an electricity supplier A are not necessarily those that will constitute the enriched UF_6, attributed to this same electricity supplier A. They could have been interchanged with those sent by electricity supplier B or C. At the end of the transformation, the enriched UF_6 is analyzed and must satisfy a single standard defining all the necessary parameters: the ASTM C996 standard.

4.3.2.3. *Separation work and undernourishment*

To measure the service provided by an enricher, the unit used is the SWU. A batch of enriched uranium is defined by two fundamental characteristics: its quantity and its grade. However, to determine the SWUs contained in this batch, a third variable is needed: the quantity of natural uranium used to fabrication this batch, or the associated tails assays content. Enrichers have the option of setting the tails assays of their facility at the desired value. This will have an impact on the quantity of UF_6 required to fabricate a given quantity of enriched UF_6 at a certain grade.

In the contracts signed with the electricity suppliers, the value of the tails assays is fixed because the supplier must deliver a fixed quantity of natural uranium in the form of UF_6 to the enricher and pay a sum calculated according to the number of SWUs established by these fixed parameters (grade, tails assays, quantity of natural UF_6, quantity of enriched UF_6). The price is therefore known in advance.

There is an optimization technique for the enricher to save natural UF_6 via the undernourishment phenomenon. It suffice to lower the tails assays of a plant compared to the ones defined in the contract signed with the electricity supplier, which has the consequence of increasing the separation work by reducing the quantity of natural uranium required for obtain the same quantity of enriched uranium with the same concentration. As a result, a part of the UF_6 supplied by the buyer is not used for the production of the sold enriched uranium but becomes the property of the enricher. This optimization varies depending on SWU market indicators, the conversion service and the cost of uranium.

4.3.2.4. *The enrichment market*

4.3.2.4.1. General view

Figure 4.22 presents the evolution of the price of enrichment since 2017.

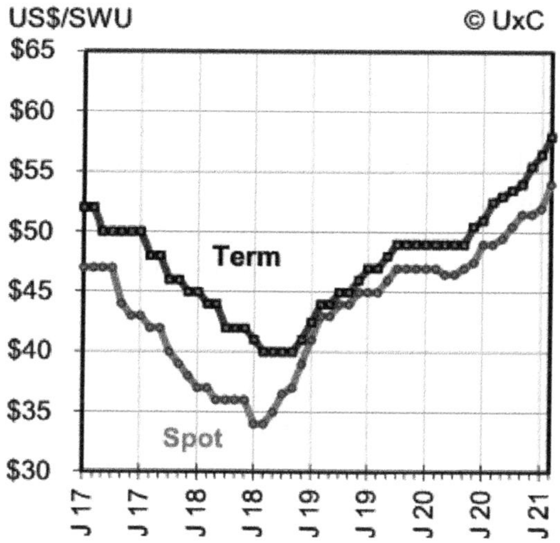

Figure 4.22. *Evolution of the enrichment price from 2017 to 2021 (source: UxC Market Outlook, 2021 update, UxC). For a color version of this figure, see www.iste.co.uk/percebois/nuclear1.zip*

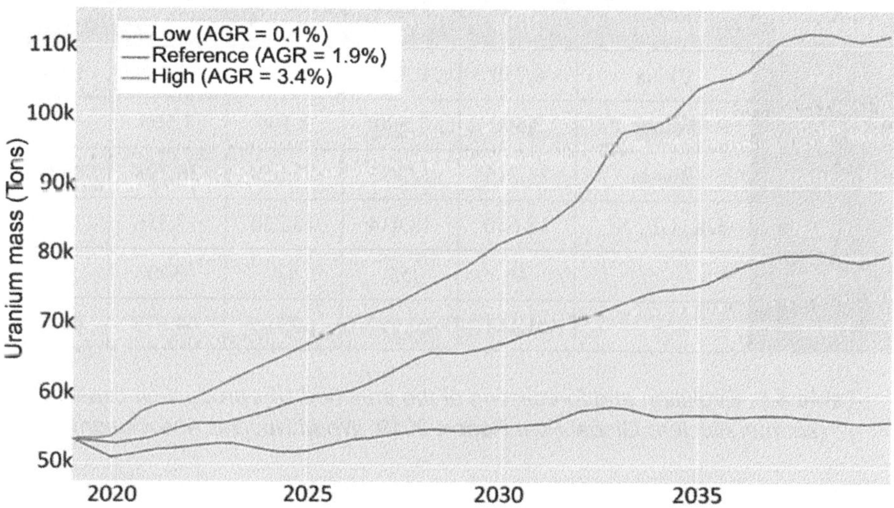

Figure 4.23. *Enrichment demand scenarios until 2040. AGR stands for annual growth rate (source: Nuclear Global Fuel Report 2019, World Nuclear Association). For a color version of this figure, see www.iste.co.uk/percebois/nuclear1.zip*

As for the conversion, the selling price of the SWU has dropped considerably over the last 10 years for the same main reasons: the consequences of the Fukushima accident, the shutdown of Japanese nuclear energy plants, the postponement of the construction of new reactors, the ongoing activity of the world's enrichment capacities and down blending (decrease in the rate of enrichment through the mix with natural or depleted uranium) of uranium resulting from the decommissioning of the American nuclear weapons lot. The price reached a maximum of \$170/SWU in 2010, before collapsing to a value around \$40–50/SWU.

4.3.2.4.2. Demand forecasting

The new projections for the conversion demand of the next 20 years predict that at the very least, the demand will stabilize (Figure 4.23).

4.3.2.4.3. Global production capacities

Despite the economic situation of the enrichment market, enrichment plants are operating at full capacity (Table 4.6) and there is no decrease in production: this is due to the high CAPEX and the low OPEX of enrichment plants.

	Production capacity (SWU)					
Operator	**Country**	**2018**	**2019**	**2020**	**2025**	**2030**
CNNC	China	6,750	6,750	6,750	13,238	19,644
Orano	France	7,500	7,500	7,500	7,500	7,500
Rosatom	Russia	28,215	27,933	27,654	26,298	25,000
Urenco	UK, GE, NL	18,600	18,414	18,230	17,336	16,487
Others	–	46	55	66	300	450
Total	–	61,111	60,652	60,199	64,673	69,081

Table 4.6. *Evolution and forecasting of the enriched UF_6 production capacity (source: Nuclear Global Fuel Report 2019, World Nuclear Association)*

Although according to average and optimistic scenarios, the demand will increase over the next few years, and there are no plans to increase production capacities. Moreover, as the construction time of a reactor is much longer than that of an enrichment plant, the market has the ability to adapt to the demand.

The uranium production market is based on three major stakeholders: Orano, Rosatom and Urenco.

Orano

Orano replaced its Eurodif plant located on the Tricastin site based on gaseous diffusion technology with the SET (*Société d'enrichissement du Tricastin*) plant, also located on the Tricastin site and based on centrifugation. In 2016, the new plant reached its nominal capacity at 7.5 SWU/year.

Orano had also planned to build an enrichment plant in the United States for the US domestic market. In 2010, Orano even obtained a $2 billion loan from the DOE for this project. However, after the Fukushima accident and the fall in the course of the SWU, the project was suspended, and definitively canceled a few years later.

Orano is now positioning itself to develop an enrichment plant to produce HALEU (High-Assay Low-Enriched Uranium, that is, uranium enriched at 19.75%) for the development of new fuel elements. Given its joint venture with Urenco, Orano has the expertise and capacities to design and operate such a plant.

Rosatom

Rosatom owns and operates four enrichment plants (Novouralsk, Angarsk, Seversk, Zelenogorsk), in partnership with the Russian company specializing in the nuclear fuel cycle TVEL. Any increases in the production capacity by optimizing current plants or by creating new ones are decided on the basis of the market. Enriched uranium has two different outlets:

– it is either used for the fabrication of fuels in the TVEL plants for the Russian domestic market;

– or it is marketed by the Russian state-owned company TENEX for the Western market. It should be noted that a restriction imposed by the United States does not enable Rosatom to sell over 20% to meet the US reactors' SWU requirements.

Enriched uranium is transported by sea via the port of Saint Petersburg for the Western market and via the port of Vostochny for the Asian market.

Urenco

Urenco has four enrichment plants, three of which are located in Europe (Capenhurst (Great Britain), Almelo (Netherlands), Gronau (Germany)) and one in the United States (New Mexico). All these plants are based on centrifugation technology for a total capacity of 18.6 SWUs.

Like Orano, Urenco is also embarking on the construction of an enrichment plant for manufacturing HALEU. This plant is located in the United States at the conventional enrichment plant site.

At the end of 2019, Urenco started increasing its depleted uranium deconversion capacity at Capenhurst, transforming depleted UF_6 into depleted U_3O_8, a more stable and less dangerous form.

CNNC

China's enrichment capacities were estimated at 6.7 MSWU, while several projects to increase this capacity are underway. Taking into account the stocks created during the last decade and the increases in production capacities, China (via CNNC), can be self-sufficient in terms of enriched uranium. Given the growth of the Chinese market, CNNC will continue its expansion strategy over the next few years.

USEC-Centrus

USEC (United States Enrichment Corporation), which became Centrus in 2014, is an American state-owned company in charge of enrichment activities for the American Federal State in order to supply American power reactors with enriched uranium. During the 1993–2013 period, USEC was also the commercial agent in charge of marketing enriched uranium from the decommissioning of atomic weapons.

In 2014, the Paducah enrichment plant, based on gaseous diffusion technology, was shut down. It used to produce 6 MSWU/year. Since then, Centrus has been working on replacing it with a centrifugation-based plant. Despite the success of its pilot, the project to build a new production plant was discontinued. However, in order to preserve the American knowledge on centrifuge technology, Centrus continues to improve its pilot.

Following a request from the DOE (Department of Energy), Centrus has also launched the development of an enrichment plant aimed at producing HALEU to be used in American research reactors.

Centrus plays a role as a trader in the enrichment market, sourcing from third parties.

GLE

This is a commercial structure bringing together GE, Hitachi and Cameco whose goal is to develop and industrialize laser-based enrichment technology. Pilots have been developed over the past 15 years but the sharp decrease in SWU prices over the past decade has considerably slowed the project, and partners have undergone restructuring following GE's intention to leave the project. The project still exists but is greatly slowed down.

Other producers

Argentina has a plant based on gaseous diffusion, which remained closed after 20 years of operation. Despite being reopened in 2015, it shut down again 3 years later in 2018. The uranium produced has never been marketed. A few exploration programs based on centrifugation or laser enrichment have been launched, without a firm schedule being established.

INB (*Indústrias Nucleares do Brasil*) has launched a program to renovate and increase the capacity of its Resende plant, which used to produce small amounts of enriched uranium in the past. The main goal is to make Brazil self-sufficient in supplying its own reactors.

For 10 years, via the JNFL company, Japan has worked on the construction of a small production unit (450 kSWU), which should be completed by 2024.

India, Iran and Pakistan have very small production units, which are not marketed. Other countries such as Canada, South Africa or Australia have conducted studies without announcing any concrete plan.

4.3.2.4.4. Market trading issues

Given its strategic nature, the enrichment market is subject to numerous commercial conflicts between the various stakeholders. These main disputes will be described in the following paragraphs.

Russian Suspension Agreement

This American measure dates from 1992 and was aimed at prohibiting the sale of uranium and enrichment services of Russian origin to the American domestic market. However, the enriched uranium from the downblending of nuclear weapons was not affected by the suspension. In 2008, some amendments to this suspension were made authorizing the import of small quantities (up to a maximum of 20% of the US domestic market) as from 2014. This restriction is still in effect today.

USEC Trade Case

From 2000 to 2019, the US Department of Commerce imposed a tax on the permanent import of enriched uranium whose enrichment was done in France, considering that the latter was subsidized by France (unlike the United States). This tax was lifted in 2019.

The Corfu Declaration

In 1994, the European Union limited the import of enriched uranium from third countries to 20% of the needs of its domestic market. However, contracts already concluded before 1994 were not affected by this restriction. Generally, the long-term contracts signed for enrichment can run for several decades.

4.3.2.4.5. Future prospects

As illustrated in the preceding paragraphs, the enrichment market currently presents excess capacities, and will continue to do so in the future, in the medium and low scenarios. However, several factors should be observed.

The first is the limited lifespan of centrifuges, many of which will expire in the next few years and will require replacement. In addition, as centrifuge technology improves rapidly, some centrifuges become obsolete, thus reducing production capacities. Given the market, some producers could be led not to replace certain centrifuges, which will consequently reduce production capacities.

The second point is the production capacity of Russia estimated at 28 SWUs, that is to say, almost half of current production capacity. However, this figure must be relativized. On the one hand, Russia uses these SWUs to re-enrich depleted uranium (*tails*), something which mechanically increases the number of SWUs required to produce 1 kg of enriched uranium. On the other hand, the production of uranium is subject to significant restrictions in Western countries, which limits the access to these markets.

Finally, the various producers in Western countries can use their capacities to re-enrich tails or to practice underfeeding (decrease the depleted uranium waste). Either way, more SWUs are needed to produce the same amount of enriched uranium. These trade-offs can be discussed depending on market conditions.

4.3.3. *Fuel design and fabrication*

This is the last step to obtain fuel assemblies. The main sub-stages are UF_6 deconversion into OU_2, the fabrication of accurately shaped ceramic pellets from incoming UO_2, the production of rigid metal frameworks for the fuel assembly and assembling the final fuel structure.

The designs of fuel assemblies are varied and fuel manufacturers specialize in certain types of fuel assemblies, associated with certain reactors. In addition, fuel assemblies also differ between reloads and first-cores. The first-core is the fuel used to start the reactor. For all of these reasons, the fuel fabrication market is not a commodity market, unlike the uranium conversion and enrichment markets: fuel assemblies (Figure 4.24) are a high-tech product, requiring high-level engineering.

Fuel manufacturers can be divided into two categories: designers who have both production and design capacities, and manufacturers who only produce fuel assemblies under the license from a designer. In general, the first-cores are obligatorily provided by the designers, whereas the reloads can be provided by the two categories of manufacturers.

Manufacturing design costs are slightly different from the other front-end components, since they include a technological component and the associated intellectual property, so that designs can be licensed with the safety authorities.

Figure 4.24. *PWR fuel assembly. For a color version of this figure, see www.iste.co.uk/percebois/nuclear1.zip*

4.3.3.1. *Operations carried out by the fuel fabrication facility*

The fuel fabrication facility receives enriched UF_6 from the electricity supplier who ordered the supply of fuel assemblies. This UF_6 can be the fruit of various enrichments and match the desired fuel design, which the fuel fabrication facility can calibrate as needed.

UF_6 is then deconverted, that is to say, transformed into UO_2 powder. This powder is generally sintered and then pressed to make UO_2 pellets. These are usually cylinders with a 10 mm diameter and a 10 mm height. After various density and geometry checks, these pellets are introduced into hollow metal cylinders to form rods which will be joined to form assemblies.

4.3.3.2. *The different fuel designs*

The main types of fuel assemblies can be categorized depending on the type of reactors, even if there are several different designs for the same type of reactor:

– LWR (light water reactor) with two subcategories:

- Pressurized water reactor (PWR), including the Russian WWER (Water-Water Energetic Reactor): this type of reactor concentrates the majority of the demand for fuel assemblies. However, there are many designs that vary the number of rods, their arrangement, position of the guiding rods, etc. The main type is the PWR17, produced by several fuel manufacturers (Westinghouse, Framatome, Kvadrat-Fuel (Russian)).

- Boiling water reactor (BWR): it consists of BWR 10x10 and BWR11x11 fuel assemblies.

– Pressurized heavy water reactor (PHWR), which uses natural uranium as fuel.

– Gas cooled reactor (GCR, especially in the UK) which uses advanced gas-cooled reactor (AGR) fuels and whose image is shown in Figure 4.25.

– RBMK (Russian reactors) that use a special type of assembly made of a central rod surrounded by two circular bundles of rods.

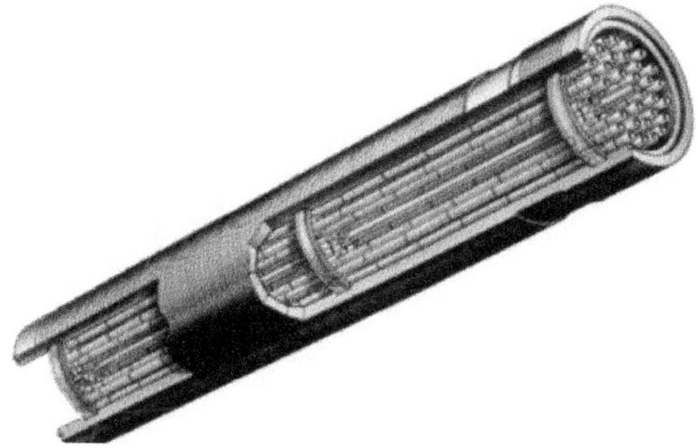

Figure 4.25. *AGR fuel assembly. For a color version of this figure, see www.iste.co.uk/percebois/nuclear1.zip*

4.3.3.3. *The demand for fuel fabrication*

This demand can be divided into two distinct categories. The first one concerns the demand for reloading reactors, whereas the other addresses first-core needs. This nuance involves two aspects: the demand predictability for reloads and the difficulty for planning first-core needs in advance. This is due to the frequent schedule changes during the construction of new reactors, which can take up to several years, and the fact that reloads can be fabricated by many suppliers, whereas first-cores can only be fabricated by the reactor's designer.

The advantage of having several qualified potential suppliers is to secure supplies of fresh fuel in the event of failure. Additionally, one same supplier can be hired, provided that it has several production sites.

For the American and European market, where very few new reactors are under construction and existing reactors are more or less reaching the end of their life, the trend in contracts is to hire a single supplier over a long period by means of supplementary agreements.

4.3.3.3.1. The manufacturing demand for fuel reloads

The evolution of demand is shown in Figure 4.26.

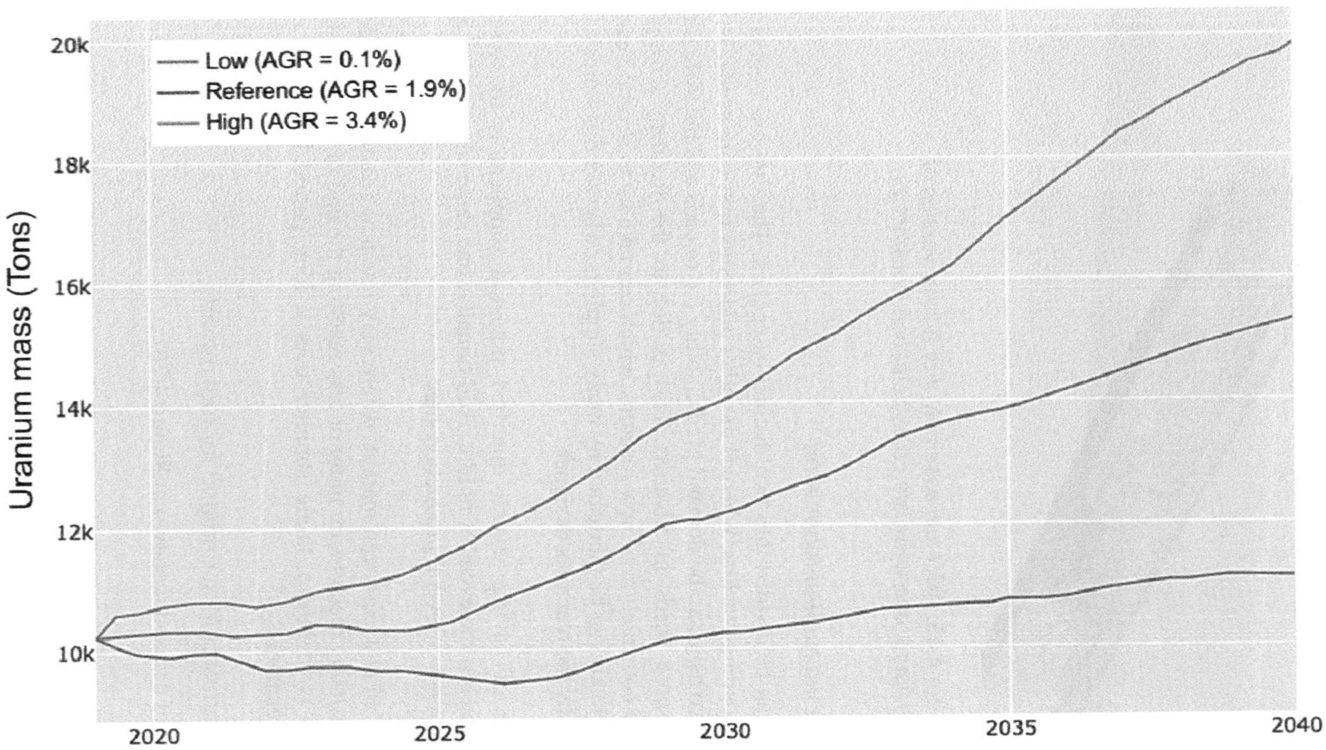

Figure 4.26. *Fuel reload demand scenarios until 2040. In this case, AGR stands for annual growth rate (source: Nuclear Global Fuel Report 2019, World Nuclear Association). For a color version of this figure, see www.iste.co.uk/percebois/nuclear1.zip*

Initially, we see a relatively stable demand which then increases, from 2025. This is explained by the diversity of situations depending on the different geographical areas:

– Europe/the United States: the market is stable (or sometimes with a decreasing trend), particularly in Europe. This is explained by the reactor shutdown (AGR, RBMK and PWR) and nuclear power phase-out from certain countries. In the American zone, it is in particular the shutdown of resource-intensive Candu-type reactors that has led to this phasing-out.

– Asia/Middle East: this is the region experiencing the highest growth, particularly with China and India bolstering the demand. This explains the strong demand growth beyond 2025.

4.3.3.3.2. The manufacturing demand for first-core fuels

Figure 4.27 illustrates this demand.

These figures are directly proportional to the number of reactors that will be built in the next 20 years. Again, demand is largely driven by the Asian market.

4.3.3.4. *Main manufacturers in the market*

Designers are the main stakeholders in the market.

4.3.3.4.1. Westinghouse

Westinghouse is a traditional market company with manufacturing capacities for all fuel types, except RBMKs. Westinghouse has three plants in Europe (Vasteras in Sweden, Springfields in England and Juzbado in Spain) and one in the United States (Columbia), each dedicated to its local market.

Metallic structures are made in the United States and supplied to European plants. Wet deconversion is done in Sweden and Columbia, whereas dry deconversion takes place in Springfields. The Juzbado plant does not include a deconversion process and directly receives UO_2 from other deconversion plants.

4.3.3.4.2. Framatome

This French company was created in 1958 to supply the French and overseas market. After its merger with Siemens Nuclear Reactor Business in 2001, it was renamed Areva NP before being sold to EDF (75.5%), Mitsubishi (19.5%) and Assystem (5%) in 2018.

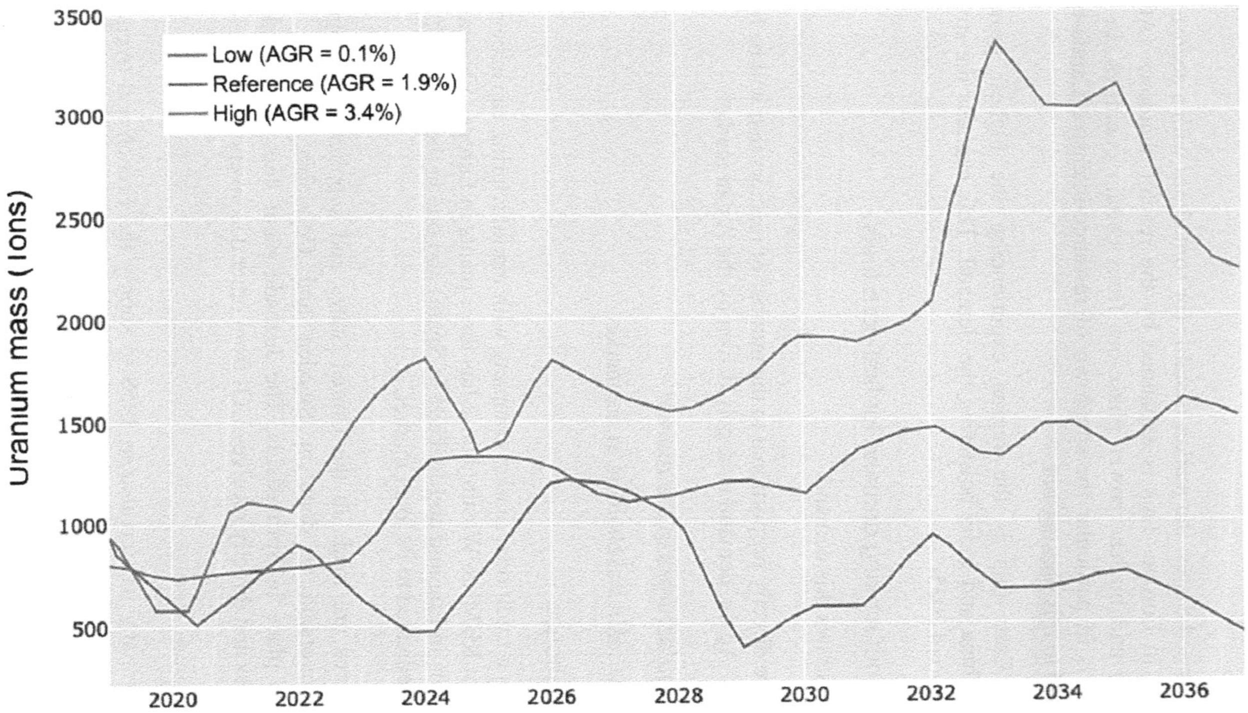

Figure 4.27. *First-core fuel demand scenarios until 2037. In this case, AGR stands for annual growth rate (source: Nuclear Global Fuel Report 2019, World Nuclear Association). For a color version of this figure, see www.iste.co.uk/percebois/nuclear1.zip*

Framatome designed the French and German reactors, including the Generation II and the EPR. Framatome has two plants in Europe (Romans in France and Lingen in Germany) and one in the United States (Richland). The three plants have deconversion, pellet manufacturing and assembly lines.

The Romans plant has a production line for power reactors and a line for research reactors. It also owns the manufacturing workshops for the metallic structures for fuel assemblies.

The Richland plant, dedicated to the American market, also has uranium recovery services and fabrication gadolinium assemblies. The Lingen plant notably fabricated the first-core for the Finnish EPR.

4.3.3.4.3. TVEL

TVEL is a Russian company that has designed RBMK and WWER reactors. TVEL has two plants in Russia (Novosibirsk and Electrostal).

Over the past few years, TVEL has diversified and has the ability to supply PWR assemblies which have already been delivered to Sweden and the United States via a partnership with GNF America. TVEL can also supply fast reactor assemblies and PWR fuel assembly components operating with reprocessed uranium.

TVEL has a deconversion technology based on the dry process and can supply the mechanical components required by the fuel assembly structures.

4.3.3.4.4. GNF

GNF is a joint venture between Hitachi and GE with a plant in Wilmington, USA, dedicated to the BWR reactor markets in the United States, but also in Europe via a joint venture with ENUSA. In Japan, the GNF-J subsidiary focuses on the Japanese market.

4.3.3.4.5. Kepco Nuclear Fuel

Kepco is a South Korean state-owned company which has a fuel fabrication plant (Daejeon) whose capacities include dry conversion, structure manufacturing and fuel assemblies.

KNF supplies the Korean domestic market as well as the United Arab Emirates.

4.3.3.4.6. CNNC

This is a Chinese state-owned company that is the only Chinese company authorized for the fabrication of nuclear fuels. CNNC has two plants in Boatou and Yibin.

Its plants supply the Chinese and Pakistani markets. CNNC has the ability to fabricate PWR, PHWR, or HTR (high-temperature reactor) fuel assemblies.

4.3.3.4.7. Mitsubichi Nuclear Fuel

Mitsubichi Nuclear Fuel (MNF) is a Japan-based company that has the capacity to design, fabricate (including deconversion) and transport fuel assemblies. MNF can also supply assemblies for BWRs. Framatome is a 5% shareholder.

4.3.3.4.8. Nuclear Fuel Industries

This is a Japanese company founded in 1974, initially developed for the Japanese market. It is equipped with all the possible capacities: design, fabrication, transport for PWR and BWR fuel assemblies.

NFI has developed fuel assemblies for Triso reactors. NFI has two plants (Osake, Tokai-Mura), each dedicated to a type of PWR or BWR fuel.

4.3.3.4.9. Enusa

This is a Spanish state-owned company, founded in 1972 and based in Salamanca. It has no deconversion capacity and produces PWR and BWR-type fuel assemblies.

4.3.3.4.10. BWXT Nuclear Energy Canada

This is a subsidiary of BWXT Canada whose purpose is the production of fuel for Candu reactors around the world. It has three plants, each dedicated to a particular unit operation: the fabrication of pellets from UO_2 in Toronto, metal fabrication in Arnprior and the final assembly in Peterborough. It also offers fuel-related services.

4.3.3.4.11. Cameco Fuel Manufacturing

Cameco is a Canadian company owning a conversion plant. It also has two plants for fuel fabrication for Candu reactors. Not only does Cameco Fuel Manufacturing supply fuel assemblies, but it is also an important component supplier for Candu reactors.

4.3.3.4.12. Ulba Metallurgical Plant

Ulba Metallurgical Plant (UMP) is a Kazakh state-owned enterprise founded in 1949. UMP owns a plant that has the capacity to fabricate pellets for nuclear energy plants not only from UF_6, but also uranium oxide, metallic uranium, yellow cakes or any type of uranium tails.

UMP supplies fuel assemblies for China (TVS) and is responsible for the construction and operation of a nuclear energy plant via a joint venture with CGN.

A few other small producers are also generally present on local markets, such as INB (Brazil), Conuar (Argentina), DAE (India) or SNN (Romania).

4.3.3.5. *The production capacity of stakeholders*

Table 4.7 shows the production capacities of the major market stakeholders.

	Fabrication facility	Locations	Conversions	Pellets	Rods/ assemblies
Brazil	INB	Resende	160	120	400
China	CJNF	Yibin	800	800	800
	CBNF	Baotou	0	0	400
	CNNFC	Baotou	200	200	200
France	Framatome-FBFC	Romans	1,800	1,400	1,400
Germany	Framatome-ANF	Lingen	800	650	650
India	DAE	Hyderabad	48	48	48
Japan	NFI (PWR)	Kumatori	0	383	284
	NFI (BWR)	Tokai-Mura	0	250	250
	Mitsubishi	Tokai-Mura	450	440	440
	GNF-J	Kurihama	0	620	630
Kazakhstan	ULBA	Ust-Kamenogorsk	0	108	0
South Korea	KEPCO NF	Daejeon	700	700	700
Russia	TVEL	Elektrostal	1,500	1,500	1,560
		Novosibirsk	450	1,200	1,200
Spain	ENUSA	Juzbado	0	500	500
Sweden	Westinghouse AB	Vasteras	787	600	600
Britain	Westinghouse	Springfields	950	600	860
United States	Framatome Inc.	Richland	1,200	1,200	1,200
	GNF	Wilmington	1,200	1,000	1,000
	Westinghouse	Columbia	1,600	1,594	2,154
Total	–	–	12,645	13,913	15,276

Table 4.7. *Changes in the demand for first-core fuels, expressed in tons of heavy metals, depending on different scenarios (source: Nuclear Global Fuel Report 2019, World Nuclear Association)*

4.3.4. *Development prospects: advanced ATF fuels*

After the Fukushima nuclear accident, the fuel development effort has focused on designs improving safety under accident conditions, accident-tolerant fuel or advanced technologies fuel (ATF).

The evolution toward long irradiation cycles should continue with these developments and a reassessment of the ^{235}U traditional maximum enrichment limits set at 5% and the maximum combustion rate set at 62 GWd/TML.

This operating strategy will reduce the number of new or fresh fuel assemblies, as well as spent fuel assemblies from the core. Increasing the fuel irradiation rate requires the use of more enriched uranium, HALEU (varying from 5 to 20%).

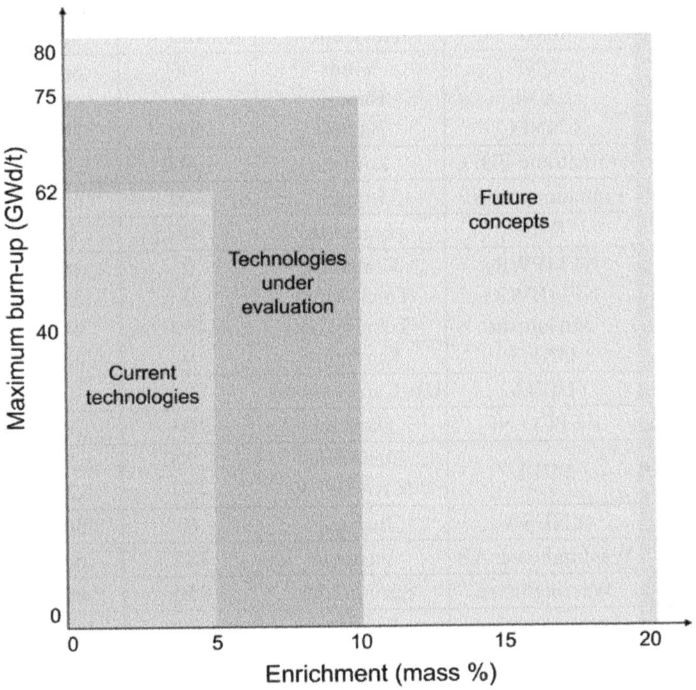

Figure 4.28. *Limits to combustion rates and fuel enrichment (Nuclear Energy Institute 2019). For a color version of this figure, see www.iste.co.uk/percebois/nuclear1.zip*

Finally, ATFs could have the potential to improve operational flexibility, enabling greater integration with renewable energies, increasing the value of the energy produced by the nuclear fleet.

These developments require a significant qualification effort depending on the deviation from the designs currently used, as well as the adaptation or development of specific front-end infrastructures. Until now, the management of advanced spent fuels has been the subject of very few studies. The impacts of advanced nuclear fuels in the cycle's back end, their stocking, transport, reprocessing and storage still have to be qualified and quantified.

Figure 4.28 illustrates the current and potential design limitations of the enrichment and irradiation rates (Nuclear Energy Institute 2019).

These developments are also associated with the development of small modular reactors (SMR). By increasing fuel enrichment to HALEU levels, with higher unit fuel costs compared to standard enrichment, a reactor could improve its load factor and reduce the average fuel supply costs per year.

4.4. References

IEA/NEA (2020). Projected costs of generating electricity. Available at: https://www.oecd-nea.org/upload/docs/application/pdf/2020-12/egc-2020_2020-12-09_18-26-46_781.pdf.

Nuclear Energy Institute (2019). White paper: The economic benefits and challenges with utilizing increased enrichment and fuel burnup for light-water reactors. Available at: https://www.nrc.gov/docs/ML2125/ML21259A191.pdf.

OECD/NEA (1994). Environmental and ethical aspects of long-lived radioactive waste disposal. Available at: https://www.oecd-nea.org/upload/docs/application/pdf/2019-12/nea120-environmental-ethical-aspects.pdf.

OECD/NEA (2013). The economics of the back end of the nuclear fuel cycle. Available at: https://www.oecd-nea.org/upload/docs/application/pdf/2019-12/7061-ebenfc.pdf.

SKB (2021). Nuclear fuel approval and ATFs in the Swedish System. Available at: https://www.nwtrb.gov/docs/default-source/meetings/2021/may/sjoland.pdf?sfvrsn=6.

World Nuclear Association (2017). Nuclear power economics and project structuring. Available at: https://www.world-nuclear.org/getmedia/84082691-786c-414f-8178-a26be866d8da/REPORT_Economics_Report_2017.pdf.aspx.

Appendix

Ore Deposits and Mining Projects

Sophie GABRIEL[1,2] and Christian POLAK[3]

[1] *I-tésé (CEA), Gif-sur-Yvette, France*

[2] *Université Paris-Saclay, Saint-Aubin, France*

[3] *Orano Mining, Paris, France*

A.1. Uranium geochemistry, a chance on Earth

Uranium is a natural element present in the Earth's crust, most often in the form of oxides or silicates, never in a metallic state. It is relatively abundant, around 2.3 g/t (or 2.3 ppm), more than tin, tungsten, precious metals or iodine. The crystallo-chemistry of uranium orients it toward differentiated magmas, granites or rhyolites, acid rocks. It finds refuge in zircons and monazites in particular.

The chemistry of uranium is very rich because it has two U^{4+} and U^{6+} cations (uranyl compounds), which are highly dependent on oxidation conditions.

The properties of uranium endow it with an active geochemistry and explain its wide dispersion, as well as its reconcentration in very high-grade deposits. Uranium can be combined with many other elements, forming a variety of minerals (249)[1,2], but only 10 of them present a mining interest: primary minerals such as uraninite (88% U), coffinite (73% U), branerite (33% U) as well as other minerals grouped

1 Available at: http://webmineral.com/chem/Chem-U.shtml#.Xx7UmKgzZaQ.

2 *Minerals & Fossils*, Special Issue No. 28, *Uraniferous Minerals*. 2009.

under the term "yellow products", often amorphous alteration minerals (gummite), which can be silicates, phosphates (autunite) or vanadates (carnotite).

Oxidation–reduction conditions are fundamental for the mobility, concentration and precipitation of uranium and, in addition to its radioactivity for detection, they are useful for its exploration and extraction.

Thorium, more abundant, 8.1 ppm, does not share this chemistry, which is reduced to a cation. Thorium is not very mobile, on the other hand. The ore deposits remain at relatively low grades and are more associated with rare-earth elements.

Uranium was identified in 1789 in a Bohemian mineral, pitchblende (the concreted facies of uraninite), by the German Martin Klaproth. It was isolated in a metallic state by the Frenchman Eugène Peligot over 50 years later in 1841.

It was in 1896 that Henri Becquerel discovered radioactivity, and in 1898, Marie and Pierre Curie isolated radium from pitchblende.

Natural uranium comprises four isotopes (234, 235, 236, 238) but only two of them, 238 (99.27%) and the fissile isotope 235 (0.72%), are predominant.

A.2. Types of ore deposits

Uranium ore deposits are classified in relation to the main fractionation processes during the geological cycle: from the mantle, to the continental crust with surface oxidation and genesis of the ore deposits.

In 2013[3], the International Atomic Energy Agency (IAEA) published a geological classification of 15 types of uranium ore deposits, listing them from the deepest to the most superficial:

- Type 1: intrusive;
- Type 2: granite-related;
- Type 3: polymetallic iron-oxide breccia complex;
- Type 4: volcanic-related;
- Type 5: metasomatite;

3 New geological classification of uranium ore deposits from the IAEA: *Geologists*, 181, June 2014.

– Type 6: metamorphite;

– Type 7: proterozoic unconformity;

– Type 8: collapse-breccia pipe;

– Type 9: sandstone;

– Type 10: paleo-quartz-pebble conglomerate;

– Type 11: surficial;

– Type 12: lignite and coal;

– Type 13: carbonate;

– Type 14: phosphate;

– Type 15: black shale.

All of the world's ore deposits are grouped together under the UDEPO database from the IAEA[4]. To date, there are over 4,100, but the classification does not take economic criteria into account. It also incorporates depleted ore deposits. The majority are sandstone deposits (Type 9 according to IAEA classification); the richest are the proterozoic unconformity deposits (Type 7) and a single deposit, the Olympic Dam (Type 3), concentrates 20% of the world's resources, with more than 2 million tons of uranium.

Most of the uranium production (80–90%) comes from Type 1, 3, 7 and 9 ore deposits, whose main mines and deposits around the world will be briefly described in the following sections.

A.2.1. *Intrusive ore deposits*

Intrusive ore deposits represent 8% of the world's production and are primarily found in two mines: Rössing and Husab, Namibia.

These are granites in which uranium is found at a few hundred ppm (despite being low grades, they represent large volumes). Uranium, carried by uraninite, is diffused in a granitic rock called alaskite. These ore deposits originated at the beginning of the Paleozoic, that is to say, 510 million years ago. Both are open mines.

4 Available at: https://infcis.iaea.org/UDEPO/Deposits.

The Rössing mine, recently acquired by the Chinese group CNNC, has produced 116 ktU since 1976 and will probably continue beyond 2030. It is the poorest mine in the world, between 0.025 and 0.035% U_3O_8.

The other mine, Husab, belonging to the Chinese group CGN, was put into production in 2017 and has become the largest uranium mine in Africa.

A.2.2. *Ore deposits associated with breccia complexes*

The only exploited deposit of this type is the Olympic Dam (BHP in South Australia), which represents 5% of the world's production. It originated 1,590 million years ago. Uraninite, coffinite and branerite are the carrier minerals.

It is above all a copper mine, where uranium is co-produced, together with a little gold and silver. It is the world's largest resource of uranium, mined underground.

Production began in 1988; over 80 ktU have been extracted to date. The grade is about 0.06% U_3O_8.

A.2.3. *Ore deposits associated with proterozoic unconformities*

Note that 20–25% of the world's production is drawn from ore deposits associated with proterozoic unconformities.

The unconformity is an interface between the rocks of the Archean basement and those of the sedimentary cover from the Proterozoic age, in this case, the uranium ore deposits. The characteristic of these ore deposits is their exceptional uranium grade (1–20% U_3O_8), an enrichment 100,000 times the grade of the Earth's crust.

Two regions provide this production, mainly Athabasca in the north of the Province of Saskatchewan in Canada and, in a smaller proportion, Kombolgie, in Australia's Northern Territory.

The ages are around 1,400–1,500 million years for Australia (Ranger mine, Energy Resources Australia, Rio Tinto group which closed in 2021) and 1,590 million years for Canada. The mineralization is essentially massive uraninite.

The circulation of saline fluids rich in uranium, interacting with reducing substrates ("graphitic cords"), gives rise to small-sized ore deposits with outstandingly rich resources: the McArthur mines (280 ktU), Cigar Lake (141 tU) operated by Cameco with the participation of Orano, or the new Arrow deposit (134 ktU) that has not yet been exploited.

These ore deposits have the highest grade value per ton in the entire mining industry.

A.2.4. *Ore deposits associated with sandstone*

Ore deposits associated with sandstone represent 50% of the world's production.

These are the most common uranium ore deposits (45%), which can be divided into two groups, those exploited by ISR (in situ recovery) or by ISL (in situ leaching) technologies.

This distinction is fundamental, because approximately 45% of the world's production uses ISR technology. Kazakhstan, the lead world producer, has multiplied its production by 12 in 20 years, exclusively relying on SRI technology. There are about 15 entities mining uranium in Kazakhstan, all of which have a stake in the Kazakh national company, Kazatomprom (the major stakeholder). Joint ventures have been signed with the Russians (Uranium One/ARMZ), the French (Orano-Katco), the Chinese (CGN-Semizbay), the Canadians (Cameco-Inkai) and the Japanese.

Next come Uzbekistan (with a national company, Navoi) and the United States with a few operations (Peninsula, Energy Fuels, Ur-Energy) in Wyoming and historically in Texas. Quasar's (a subsidiary of General Atomic) Four Mile mine in South Australia was the only Australian SRI mine in operation in 2020. The Chinese company CNNC uses ISR in Xinjiang and Inner Mongolia. ARMZ operates two ISR mines, Khiagda and Dalur, in Russia.

The uranium solutions resulting from the leaching of granite massifs circulate and migrate toward the permeable layers (sands, sandstones) of the sedimentary basins and come across smaller zones into which uranium precipitates. It thus forms the so-called roll-front structures. ISR leaching amounts to prompting the recirculation of uranium by injecting acid or basic solutions, if limestone is present. It is essential for the uranium mineralized layers to be confined by impermeable layers (clays) ir order to avoid any loss or dilution. The solutions are then pumped to the surface and processed.

Grades are often lower than 0.1% U_3O_8, but the ease of exploitation with reduced investments (no quarry or galleries, only migration surveys) makes this technology extremely attractive with the lowest production costs on the market.

The age of roll fronts is often different from the age of the supporting sedimentary layer. According to tectonic movements, ages are rather recent: 20–25 million years in Xinjiang and 2–20 million years in the Kazakh basins of Shu-Syrasu and Syr-Daria. Roll fronts are dynamic formations that always evolve if a hydraulic gradient is present.

In Kazakhstan, uranium comes from the Karatau ridge, or is generated by the oxidation of minerals present in the sands and carrying uranium (monazite, ilmenite).

For the other sandstone ore deposits, their resources can be significant: in Niger, the two Somaïr and Cominak mines operated by Orano, set up in 1971 and 1978 respectively, have produced 153 ktU. Cominak ceased its activity in 2021.

The primary sandstone mineralizations, due to the leaching of the granitic massifs of the Aïr in Niger, are quite young (around 100–135 million years). The ore deposits appear in Jurassic and carboniferous formations.

The characteristics of the ore deposits are multiple and greatly vary from one deposit to another: grade, depth, size, hardness of the rock or cover, thickness of the mineralized zones, presence of carbonates or organic matter, other recoverable (Cu, Au) or prejudicial (Mo, Zr) elements. Finally, to transform an ore deposit into technically and economically exploitable resources or reserves, it is the whole setting that must be evaluated, in accordance with a precise nomenclature.

List of Authors

Sylvain DAVID
IJCLab Orsay, CNRS
IN2P3
France

Jean-Guy DEVEZEAUX DE LAVERGNE
Formerly I-tésé (CEA)
Gif-sur-Yvette
France

Cécile EVANS
Orano
Châtillon
France

Sophie GABRIEL
I-tésé (CEA)
Gif-sur-Yvette
and
Université Paris-Saclay
Saint-Aubin
France

Frédéric GARCIAS
IAE Lille
and
Université de Lille
France

Daniel IRACANE
CEA
Saclay
France

Jacques PERCEBOIS
CREDEN, CNRS ART-Dev
Université de Montpellier
France

Christian POLAK
Orano Mining
Paris
France

Nicolas THIOLLIÈRE
Subatech
IMT Atlantique
Nantes
France

Stéphanie TILLEMENT
LEMNA
IMT Atlantique
Nantes
France

Philippe VALBUENA
Orano
Châtillon
France

Index

C, D, E

F, H, J

L, N, P

R, S, U, W

Summary of Volume 2

Chapter 5. Prospective Scenarios from the Present to 2050
Patrick CRIQUI, Rodica LOISEL and Lionel LEMIALE

Appendix. The Dismantling of Nuclear Facilities
Jacques PERCEBOIS and Nicolas THIOLLIÈRE

Printed and bound by CPI Group (UK) Ltd, Croydon, CR0 4YY
13/09/2026
14960997-0003